Samuel Octavus Gray

British Sea-Weeds

An introduction to the study of the marine Algae of Great Britain, Ireland ad the

Channel Islands

Samuel Octavus Gray

British Sea-Weeds
An introduction to the study of the marine Algae of Great Britain, Ireland ad the Channel Islands

ISBN/EAN: 9783337325817

Printed in Europe, USA, Canada, Australia, Japan

Cover: Foto ©berggeist007 / pixelio.de

More available books at **www.hansebooks.com**

BRITISH SEA-WEEDS:

AN

INTRODUCTION TO THE STUDY

OF THE

MARINE ALGÆ

OF

Great Britain, Ireland, and the Channel Islands.

BY

SAMUEL OCTAVUS GRAY.

"Thus learn we Bounty's lore
Along the unbounded shore."—*Keble*.

LONDON:
L. REEVE & CO., 5, HENRIETTA STREET, COVENT GARDEN.
1867.

TO

JOHN EDWARD GRAY, F.R.S.,
V.P.Z.S., F.L.S., ETC. ETC. ETC.,

AND TO HIS WIFE

MARIA EMMA GRAY,

THIS LITTLE WORK IS AFFECTIONATELY INSCRIBED

BY THEIR GRATEFUL NEPHEW

S. O. GRAY.

PREFACE.

THE study of Sea-weeds,—Phycology, as it is technically termed,—was formerly a comparatively neglected science, and in consequence there was scope for more rapid progress in it than in the better known branches of botany. Accordingly, in the interval that has elapsed since the publication of Dr. Harvey's admirable 'Phycologia Britannica,' many very important changes have been made in the classification and nomenclature of Seaweeds, particularly of those of the Red series. These changes are recorded, chiefly, in expensive works, such as Harvey's 'Nereis Boreali-Americana' and 'Phycologia Australica,' etc., in scientific periodicals, or in the foreign publications of Agardh, Kützing, and others; and have not, that I am aware, been introduced into any easily accessible English work, with the exception of Dr. Gray's 'Handbook of Water-weeds.' In the following pages I have endeavoured to supply the want that appeared to me to exist,—to offer to the more or less advanced student a manual of British Sea-weeds, based on the most recent scientific research; and to the beginner an introduction to the study, and a guide to his early wanderings on the shore among the rocks and tide-pools. I have freely used all the works mentioned above, particularly the 'Nereis Boreali-Americana' and 'Phycologia Britannica,' and have endeavoured to condense the facts that they contain into the most intelligible, least technical, and briefest language that a strict regard for scientific accuracy would permit. I have, also, recorded with diffidence the results of my own experience and study, and have collected all the information that I could obtain from friends and fellow-students.

I have embodied in the text my acknowledgments to those persons who have kindly furnished me with specimens or information; I must here thank those from whom I have received more general assistance. I am most deeply indebted to my uncle, Dr. John Edward Gray. His early instruction and great example first led me to take delight in the study of botany; his recommendation obtained for me this opportunity to use the knowledge that he had induced me to acquire; his library has supplied me with the costly scientific works to which I have had to refer; and the material in his 'Handbook of Water-weeds' and his manuscript notes have been freely placed at my disposal. For all these benefits, and for his valuable counsel, which has been, during the preparation of these pages, as it has been throughout my life, my ever-ready resource on any occasion of doubt or difficulty, I am deeply grateful. To Mrs. Gray, also, I owe many thanks, for the advantages that I have derived from the study of the authentic specimens of rare and doubtful species which are contained in her rich collections. I venture to mention in this place that in addition to her other labours of love in the service of science, this talented lady has recently supplied to many large public and private schools, in various parts of the kingdom, sets of named specimens of British Sea-weeds and Ferns, in the hope that she may thereby promote a taste for those pursuits from which she herself has derived so much pleasure.

Both during the preparation of the work, and while it has been passing through the press, my wife has constantly helped me as an amanuensis, and with advice and criticism; and for the patience, perseverance, and ability with which she has done so, I offer her my loving thanks.

I desire to add a few words "in memoriam" of Dr. Harvey and of Mr. Lovell Reeve. The recent death of the former, at a comparatively early age, has robbed botanical science of one of her most distinguished votaries. His splendid works, more especially those which had reference to marine Algæ, deserve the deepest gratitude of every student of phycology, and will form for him a fitting and imperishable monument.

It was at the request of Mr. Reeve, that this work was undertaken, and he has died during its progress. Both as an author and as a publisher he has left science deeply his debtor. His labours in the former capacity are well known and recognized; those in the latter, though not so ambitious, were scarcely less useful. He strove to divest scientific books of unnecessary technicalities, and to introduce correct teaching into popular works, so as to make them the fit commencement of the pursuit of real science. The many excellent publications bearing his monogram are the best evidence of the success that he attained.

In the humble hope that my feeble efforts may not disgrace that impress, nor be found altogether unworthy of the cause in which they have been exerted, I commend the result of my labours to the tender treatment of the indulgent reader.

S. O. G.

Dalston Rise, 31*st August*, 1867.

CONTENTS.

	PAGE
SYSTEMATIC LIST OF ORDERS, GENERA, AND SPECIES	xvii
CHAPTER I.—THE POSITION OF SEA-WEEDS IN THE VEGETABLE KINGDOM; THEIR STRUCTURE, ETC.	1
CHAPTER II.—ON THE COLOUR, DISTRIBUTION, ETC., OF SEA-WEEDS	16
CHAPTER III.—ON COLLECTING AND PRESERVING SEA-WEEDS, THEIR USES, ETC.	24
OLIVE-COLOURED SEA-WEEDS.—MELANOSPERMEÆ	36
Fucaceæ	36
Sporochnaceæ	46
Laminariaceæ	51
Dictyotaceæ	55
Chordariaceæ	66
Ectocarpaceæ	75
RED SEA-WEEDS.—RHODOSPERMEÆ	87
Rhodomelaceæ	87
Laurenciaceæ	118
Corallinaceæ	125
Hapalidiaceæ	133
Sphærococcoideæ	134
Gelidiaceæ	149
Spongiocarpeæ	151

	PAGE
Squamariæ	152
Helminthocladiœ	156
Wrangeliaceæ	160
Rhodymeniaceæ	163
Cryptonemiaccæ	173
Spyridiaceœ	197
Ceramiaceæ	199

GRASS-GREEN SEA OR FRESHWATER WEEDS.—CHLOROSPER-
MEÆ 241

Siphonaceæ	241
Ulvaceæ	247
Confervaceæ	258
Oscillatoriaceæ	282
Nostochineæ	294
Bulbochætaceæ	297

EXPLANATION OF SCIENTIFIC TERMS 298
INDEX 307

LIST OF PLATES.

PLATE I.
1. Sargassum vulgare.
   1 a, air-vessel.
   1 b, spore-receptacles.
2. Halidrys siliquosa.
   2 a, section of receptacle, showing spores.
3. Pycnophycus tuberculatus.
   3 a, section of receptacle, showing spores.
4. Leathesia tuberiformis.
   4 a, filaments with spores.

PLATE II.
1. Fucus vesiculosus.
   1 a, section of receptacle, showing spores.
2. Asperococcus Turneri.
   2 a, part of frond, showing spore-clusters.
3. Laurencia cæspitosa.
   3 a, tip of branch, showing tetraspores.

PLATE III.
1. Fucus nodosus.
   1 a, segment of a spore-receptacle.
2. Fucus anceps.
   2 a, pointed spore-receptacle.
   2 b, branchlet with antheridia.
3. Haliseris polypodioides.
   3 a, portion of frond with sorus.
4. Sphacelaria filicina.
   4 a, a pinna.

PLATE IV.
1. Fucus canaliculatus.
   1 a, part of spore-receptacle.
   1 b, section of spore-receptacle.
2. Delesseria ruscifolia.
   2 a, tip of leaf with spore-conceptacle.
3. Schizymenia edulis.
   3 a, section of frond, showing spore-clusters.

PLATE V.
1. Alaria esculenta.
   1 a, section of part of a sorus.
2. Laminaria saccharina.
   2 a, slice of frond.

LIST OF PLATES.

PLATE VI.

1. Laminaria Phyllitis.
   1 a, slice of frond.
2. Lomentaria ovalis.
   2 a, a branchlet with spore-conceptacles.
   2 b, a branchlet with tetraspores.
3. Porphyra vulgaris.
   3 a, vertical section of frond.

PLATE VII.

1. Chorda lomentaria.
   1 a, transverse section of part of frond.
2. Padina pavonia.
   2 a, recurved margin.
   2 b, fringe.
   2 c, young sorus.
   2 d, old sorus.
3. Dictyota dichotoma.
   3 a, sorus.
4. Elachista fucicola.
   4 a, branched thread of tubercle, with spore.
5. Melobesia polymorpha.
   5 a, portion of frond, showing spore-conceptacles.

PLATE VIII.

1. Odonthalia dentata.
   1 a, branchlet with spore-conceptacles.
   1 b, branchlet with stichidia.
2. Nitophyllum punctatum.
   2 a, sorus.
3. Catenella opuntia.
   3 a, fronds.
   3 b, branch with spore-conceptacle.

PLATE IX.

1. Polysiphonia parasitica.
   1 a, branchlet with spores.
   2 b, branchlet with tetraspores.
2. Gelidium corneum.
   2 a, branchlet with spores.
   2 b, branchlet with tetraspores.
3. Halymenia ligulata.
   3 a, section of frond with spores.
4. Griffithsia corallina.
   4 a, part of branch with tetraspores.
   4 b, spore-clusters.

PLATE X.

1. Bonnemaisonia asparagoides.
   1 a, branch with spore-conceptacles.
2. Wrangelia multifida.
   2 a, part of a branch.
   2 b, tetraspores.
3. Callophyllis laciniata.
   3 a, spore-conceptacles.

PLATE XI.

1. Nitophyllum laceratum.
   1 a, marginal processes with tetraspores.
2. Chylocladia articulata.
   2 a, part of a branch with spores.
3. Gloiosiphonia capillaris.
   3 a, branchlet with fructification.
4. Callithamnion plumula.
   4 a, spore-clusters.

PLATE XII.

1. Sphærococcus coronopifolius.
   1 a, branchlet with spores.
2. Maugeria sanguinea.
   2 a, midrib with spore-leaflets.

3. Ptilota plumosa.
   3 a, comb-like branchlet.
   3 b, spore-clusters.

PLATE XIII.
1. Desmarestia aculeata.
   1 a, branchlet of a young frond.
2. Cladostephus verticillatus.
   2 a, whorls of branchlets.
   2 b, branchlets.
   2 c, branchlets with spores.
3. Ulva latissima.
   3 a, cellules of upper layer of frond.

PLATE XIV.
1. Sporochnus pendunculatus.
   1 a, a mature receptacle.
2. Codium bursa.
   2 a, fibres of frond.
3. Enteromorpha intestinalis.
   3 a, small portion of frond.
4. Calothrix confervicola (*on Ceramium rubrum*).
   4 a, proliferous thread.
   4 b, spores.
   4 c, portion of thread.

PLATE XV.
1. Striaria attenuata.
   1 a, part of a branch with spores.
2. Cladophora laetevirens.
   2 a, part of a branch.
3. Cladophora lanosa.
   3 a, part of a branch.
4. Chætomorpha Melagonium.
   4 a, cells of thread.

PLATE XVI.
1. Bryopsis plumosa.
   1 a, a plumule.
2. Enteromorpha compressa.
   2 a, part of a frond.
3. Cladophora falcata.
   3 a, branch.

SYSTEMATIC LIST

OF THE

ORDERS, GENERA, AND SPECIES.

[In this list *A.* is an abbreviation for Agardh, *J. A.* for J. Agardh, *B.* for Berkeley, *G.* for Greville, *J. E. G.* for Dr. J. E. Gray, *H.* for Harvey, *K.* for Klitzing, *L.* for Linnæus, *La.* for Lamouroux, *Lb.* for Lyngbye, *T.* for Turner, and *P. B.* for Phycologia Britannica.]

Olive-coloured Sea-weeds. Melanospermeæ. p. 36

I. **Fucaceæ.** 36
  I. SARGASSUM, *A.* 36
1. vulgare, *A.*
2. bacciferum, *A.*

  II. HALIDRYS, *Lb.* 39
3. siliquosa, *Lb.*

  III. CYSTOSEIRA, *A.*
4. ericoides, *A.*
5. granulata, *A.*
6. barbata, *A.*
7. fœniculacea, *G.*
8. fibrosa, *A.*

  IV. PYCNOPHYCUS, *K.*
9. tuberculatus, *K.*

  V. FUCUS, *L.*
10. vesiculosus, *L.*
11. ceranoides, *L.*
12. serratus, *L.*
13. nodosus, *L.*

14. Mackaii, *T.*
15. canaliculatus, *L.*
16. anceps, *H. and Ward.*

  VI. HIMANTHALIA, *Lb.*
17. lorea, *Lb.*

II. **Sporochnaceæ.**

  VII. DESMARESTIA, *La.*
18. ligulata, *La.*
19. pinnatinervia, *Montagne.*
20. aculeata, *La.*
21. viridis, *La.*

  VIII. ARTHROCLADIA, *Duby.*
22. villosa, *Duby.*

  IX. SPOROCHNUS, *A.*
23. pedunculatus, *A.*

  X. CARPOMITRA, *K.*
24. Cabreræ, *K.*

III. **Laminariaceæ.** 51
  XI. ALARIA, *G.*
25. esculenta, *G.*

  XII. LAMINARIA, *La.*
26. digitata, *La.*
    v. stenophylla, *K.*
27. bulbosa, *La.*
28. longicruris, *De la Pylaie.*
29. saccharina, *La.*
30. Phyllitis, *La.*
31. fascia, *A.*

  XIII. CHORDA, *Stackh.*
32. filum, *La.*
33. lomentaria, *Lb.*

IV. **Dictyotaceæ.** 58
  XIV. CUTLERIA, *G.*
34. multifida, *G.*

b

XV. Haliseris, G.
35. polypodioides, A.

XVI. Padina, Adanson.
36. pavonia, La.

XVII. Zonaria, A.
37. collaris, A.
38. parvula, G.

XVIII. Taonia, J. A.
39. atomaria, J. A.
(*Dictyota atomaria*, P.B.)

XIX. Dictyota, La.
40. dichotoma, La.
   v. intricata, G.

XX. Stilophora, J. A.
41. rhizodes, J. A.
42. Lyngbyæi, J. A.

XXI. Dictyosiphon, G.
43. fœniculaceus, G.

XXII. Striaria, G.
44. attenuata, G.

XXIII. Punctaria, G.
45. latifolia, G.
46. plantaginea, G.
47. tenuissima, G.

XXIV. Asperococcus, La.
48. compressus, *Griffiths*.
49. Turneri, *Hooker*.
50. echinatus, G.

XXV. Litosiphon, H.
51. pusillus, H.
52. laminariæ, H.

V. Chordariaceæ.
XXVI. Chordaria, A.
53. flagelliformis, A.
54. divaricata, A.

XXVII. Mesogloia, A.
55. vermicularis, A.
56. Griffithsiana, G.
57. virescens, *Carmichael*.

XXVIII. Leathesia, S. F. Gray.
58. tuberiformis, S. F. Gr.
59. crispa, H.
60. Berkeleyi, H.

XXIX. Ralfsia, B.
61. verrucosa, *Areschoug*.
(*Ralfsia deusta*, P. B.)

XXX. Elachista, *Fries*.
62. fucicola, *Fries*.
63. flaccida, *Areschoug*.
64. stellulata, *Griffiths*.
65. scutulata, *Duby*.
66. curta, *Areschoug*.
67. pulvinata, H.
68. velutina, *Fries*.
69. Grevillei, H.

XXXI. Myrionema, G.
70. strangulans, G.
71. Lechlancherii, H.
72. punctiforme, H.
73. clavatum, *Carmichael*.

VI. Ectocarpaceæ.
XXXII. Cladostephus, A.
74. verticillatus, A.
75. spongiosus, A.

XXXIII. Sphacelaria, Lb.
76. filicina, A.
77. scoparia, Lb.
78. plumosa, Lb.
79. cirrhosa, A.
80. fusca, A.
81. radicans, H.
82. racemosa, G.

XXXIV. Ectocarpus, Lb.
83. siliculosus, Lb.
84. amphibius, H.
85. fenestratus, B.
86. fasciculatus, H.
87. Hincksiæ, H.
88. tomentosus, Lb.
89. crinitus, *Carmichael*.
90. pusillus, *Griffiths*.
91. distortus, *Carmichael*.
92. Landsburgii, H.
93. litoralis, Lb.
94. longifructus, H.
95. granulosus, A.
96. sphærophorus, *Carm*.
97. brachiatus, H.
98. Mertensii, A.

XXXV. Myriotrichia, H.
99. clavæformis, H.
100. filiformis, H.

Red Sea-weeds. Rhodospermeæ.

VII. Rhodomelaceæ.
XXXVI. Odonthalia, Lb.
101. dentata, Lb.

XXXVII. Chondria, A.
(*Laurencia*, P. B.)
102. dasyphylla, A.

v. squarrosa, H.
103. tenuissima, A.

XXXVIII. Rhodomela, A.
104. lycopodioides, A.
105. subfusca, A.

XXXIX. Bostrychia, *Montagne*.
106. scorpioides, *Montagne*.

XL. Rytiphlœa, A.
107. pinastroides, A.
108. complanata, A.

109. thuyoides, *H.*
110. fruticulosa, *H.*

XLI. POLYSIPHONIA, *G.*
111. Brodiæi, *G.*
112. nigrescens, *G.*
113. atro-rubescens, *G.*
114. subulifera, *A.*
115. obscura, *J. A.*
116. parasitica, *G.*
117. variegata, *A.*
118. furcellata, *H.*
119. urceolata, *G.*
120. formosa, *Suhr.*
121. fibrata, *H.*
122. pulvinata, *Sprengel.*
123. Griffithsiana, *H.*
124. spinulosa, *G.*
125. Richardsoni, *Hook.*
126. elongella, *H.*
127. elongata, *G.*
128. violacea, *G.*
129. fibrillosa, *G.*
130. fastigiata, *G.*
131. byssoides, *G.*

XLII. DASYA, *A.*
132. coccinea, *A.*
133. ocellata, *H.*
134. arbuscula, *A.*
135. venusta, *H.*
136. punicea, *H.*
137. Cattlowiæ, *H.*

VIII. Laurenciaceæ.
XLIII. BONNEMAISONIA, *A.*
138. asparagoides, *A.*

XLIV. LAURENCIA, *La.*
139. pinnatifida, *La.*
140. cæspitosa, *Lu.*
141. obtusa, *La.*

XLV. CHAMPIA, *La.*
(*Chylocladia,* P. B.)
142. parvula, *H.*

XLVI. LOMENTARIA, *Lb.*
(*Chylocladia,* P. B.)
143. kaliformis, *Gaill.*
144. reflexa, *Chauvin.*
145. ovalis, *Endlicher.*

IX. Corallinaceæ.
XLVII. CORALLINA, *L.*
146. officinalis, *L.*
147. squamata, *Parkinson.*

XLVIII. JANIA, *La.*
148. rubens, *La.*
149. corniculata, *La.*

XLIX. MELOBESIA, *La.*
150. calcarea, *Ellis.*
151. fasciculata, *H.*
152. polymorpha, *L.*
153. lichenoides, *Borlase.*
154. agariciformis, *H.*
155. membranacea, *La.*
156. farinosa, *La.*
157. verrucata, *La.*
158. pustulata, *La.*

X. Hapalidiaceæ.
L. HAPALIDIUM, *K.*
159. phyllactidium, *K.*
(*Lithocystis Allmanni,* P. B.)

XI. Sphærococcoideæ.
LI. DELESSERIA, *La.*
160. sinuosa, *La.*
161. alata, *La.*
162. angustissima, *Griffiths.*
163. Hypoglossum, *A.*
164. ruscifolia, *La.*

LII. NITOPHYLLUM, *G.*
165. punctatum, *G.*
  v. β. ocellatum, *H.*
  γ. crispatum, *H.*
  δ. Pollexfenii, *H.*
  ε. fimbriatum, *H.*

166. Hilliæ, *G.*
167. Bonnemaisonii, *G.*
168. Gmelini, *G.*
169. laceratum, *G.*
170. uncinatum, *J. E. G.*
171. versicolor, *H.*

LIII. CALLIBLEPHARIS, *K.*
(*Rhodymenia,* P. B.)
172. ciliata, *K.*
173. jubata, *K.*

LIV. GRACILARIA, *G.*
174. multipartita, *J. A.*
175. compressa, *G.*
176. confervoides, *G.*

LV. SPHÆROCOCCUS, *Stackhouse.*
177. coronopifolius, *A.*

XII. Gelidiaceæ.
LVI. GELIDIUM, *La.*
178. corneum, *La.*
  v. β. sesquipedale, *G.*
  γ. pinnatum, *G.*
  δ. uniforme, *T.*
  ε. capillaceum, *T.*
  ζ. latifolium, *G.*
  η. confertum, *G.*
  θ. flexuosum, *H.*
  ι. aculeatum, *G.*
  κ. abnorme, *G.*
  λ. pulchellum, *T.*
  μ. claviferum, *G.*
  ν. clavatum, *G.*
  o. crinale, *G.*

XIII. Spongiocarpeæ.
LVII. POLYIDES, *A.*
179. rotundus, *G.*

XIV. Squamariæ.
LVIII. PEYSSONELIA, *Decaisne.*
180. Dubyi, *Crouan.*

LIX. HILDENBRANDTIA, *Nardo.*
181. rubra, *Meneghini.*

LX. PETROCELIS, *J. A.*
182. Cruenta, *J. A.*
(*Cruoria pellita*, P. B.)

LXI. CRUORIA, *Fries.*
183. pellita, *Fries.*
184. adhærens, *Crouan.*

LXII. ACTINOCOCCUS, *K.*
185. Hennedyi, *H.*

XV. **Helminthocla-
diæ.**

LXIII. HELMINTHORA, *J. A.*
(*Dudresnaia*, P. B.)
186. divaricata, *J. A.*

LXIV. NEMALION, *Duby.*
187. multifidum, *J. A.*

LXV. HELMINTHOCLA-
DIA, *J. A.*
(*Nemaleon*, P. B.)
188. purpurea, *J. A.*

LXVI. SCINAIA, *Bivona.*
(*Ginnania*, P. B.)
189. furcellata, *Bivona.*

XVI. **Wrangeliaceæ.**

LXVII. WRANGELIA, *A.*
190. multifida, *J. A.*

LXVIII. NACCARIA, *Endl.*
191. Wiggii, *Endlicher.*

LXIX. ATRACTOPHORA, *J. E. G.*
192. hypnoides, *J. E. G.*

XVII. **Rhodymenia-
ceæ.**

LXX. MATGERIA, *S. O. G.*
(*Delesseria*, P. B.)
193. sanguinea, *S. O. G.*

LXXI. RHODYMENIA, *G.*
194. palmata, *G.*
v. β. marginifera, *H.*
γ. simplex, *H.*
δ. Sarniensis, *H.*
ε. sobolifera, *H.*
195. palmetta, *G.*

LXXII. EUTHORA, *J. A.*
(*Rhodymenia*, P. B.)
196. cristata, *J. A.*

LXXIII. RHODYPHYL-
LIS, *K.*
(*Rhodymenia*, P. B.)
197. bifida, *K.*
198. appendiculata, *J. E. G.*

LXXIV. PLOCAMIUM, *La.*
199. coccineum, *Lb.*

LXXV. CORDYLECLA-
DIA, *J. A.*
(*Gracilaria*, P. B.)
200. erecta, *J. A.*

XVIII. **Cryptonemi-
aceæ.**

LXXVI. STENOGRAMMA, *H.*
201. interrupta, *Montagne.*

LXXVII. PHYLLO-
PHORA, *G.*
202. Brodiæi, *J. A.*
203. rubens, *G.*
204. membranifolia, *J. A.*
205. palmettoides, *J. A.*

LXXVIII. GYMNOGON-
GRUS, *Mart.*
206. Griffithsiæ, *J. A.*
207. Norvegicus, *J. A.*
(*Chondrus Norvegicus*, P. B.)

LXXIX. AHNFELTIA, *J. A.*
(*Gymnogongrus*, P. B.)
208. plicata, *Fries.*

LXXX. CYSTOCLONIUM, *K.*
(*Hypnæa*, P. B.)
209. purpurascens, *K.*

LXXXI. CALLOPHYL-
LIS, *K.*
(*Rhodymenia*, P. B.)
210. laciniata, *K.*

LXXXII. KALLYMENIA, *J. A.*
211. reniformis, *J. A.*
212. microphylla, *J. A.*

LXXXIII. GIGARTINA, *La.*
213. acicularis, *La.*
214. pistillata, *La.*
215. Teedii, *La.*
216. mamillosa, *J. A.*

LXXXIV. CHONDRUS, *Stackh.*
217. crispus, *Lb.*

LXXXV. CHYLOCLADIA, *J. A.*
218. articulata, *G.*
219. clavellosa, *J. A.*
(*Chrysymenia c.*, P. B.)
220. rosea, *H.*
(*Chrysymenia r.*, P. B.)

LXXXVI. HALYMENIA, *A.*
221. ligulata, *A.*

LXXXVII. Furcel-
  laria, *La.*
222. fastigiata, *Lb.*

LXXXVIII. Gratelou-
  pia, *J.*
223. filicina, *A.*

LXXXIX. Schizyme-
  nia, *J. A.*
224. edulis, *Stackhouse.*
  (*Iridæa edulis*, P. B.)
225. Dubyi, *Chauvin.*
  (*Kallymenia Dubyi*, P.B.)

XC. Catenella, *G.*
226. opuntia, *G.*

XCI. Gloiosiphonia, *Carmichael.*
227. capillaris, *Carm.*

XCII. Dumontia, *La.*
228. filiformis, *G.*

XIX. Spyridiaceæ.

XCIII. Spyridia, *H.*
229. filamentosa, *H.*

XX. Ceramiaceæ.
  200
XCIV. Microcladia, *G.*
230. glandulosa, *G.*

XCV. Ceramium, *Lb.*
231. rubrum, *A.*
  v. decurrens, *A.*
  (*C. decurrens*, P. B.)
  proliferum, *A.*
  (*C. botryocarpum*, P. B.)
  secundatum, *A.*
  pedicellatum, *A.*
232. diaphanum, *Roth.*
233. Deslongchampsii, *Chauvin.*
234. tenuissimum, *Lb.*
  (*C. nodosum*, P. B.)
235. gracillimum, *Grif.*
  § H.
236. strictum, *K.*
237. fastigiatum, *H.*
238. echionotum, *J. A.*
239. acanthonotum, *Carmichael.*
240. ciliatum, *Ducluz.*
241. flabelligerum, *J. A.*

XCVI. Ptilota, *A.*
242. plumosa, *A.*
243. elegans, *Bonnemaison.*
  (*P. sericea*, P. B.)

XCVII. Dudresnaia, *Bonnemaison.*
244. coccinea, *Bonnemaison.*

XCVIII. Crouania, *J.A.*
245. attenuata, *J. A.*

XCIX. Halurus, *K.*
  (*Griffithsia*, P. B.)
246. equisetifolius, *K.*
  v simplicifilum
  (*Griffithsia simplicifilum*, P. B.)

C. Griffithsia, *A.*
247. setacea, *A.*
248. secundiflora, *J. A.*
249. corallina, *A.*
250. Devoniensis, *H.*
251. barbata, *A.*
  CI. Seirospora, *H.*
252. Griffithsiana, *H.*

CII. Corynospora, *J. A.*
  (*Callithamnion*, P. B.)
253. pedicellata, *J. A.*

CIII. Callithamnion, *Lb.*
254. arbuscula, *J. A.*
255. spongiosum, *H.*
256. Brodiæi, *H.*
257. tetragonum, *A.*
  v. β. brachiatum, *H.*
  (*C. brachiatum*, P. B.)
258. tetricum, *A.*
259. Hookeri, *A.*
260. fasciculatum, *H.*
261. Borreri, *A.*
262. polyspermum, *A.*
263. tripinnatum, *A.*
264. affine, *H.*
265. thuyoideum, *A.*
266. gracillimum, *A.*
267. corymbosum, *A.*
268. byssoideum, *Arnott.*
269. interruptum, *A.*
270. roseum, *Lb.*
271. floccosum, *A.*
272. plumula, *Lb.*
273. cruciatum, *A.*
274. pluma, *A.*
275. Turneri, *A.*
276. barbatum, *A.*
277. floridulum, *A.*
278. Rothii, *Lb.*
279. mesocarpum, *Carm.*
280. Daviesii, *Lb.*
281. virgatulum, *H.*
282. sparsum, *H.*

Grass-green Sea or Freshwater Weeds. Chlorospermeæ. LA\

XXI. Siphonaceæ.
CIV. Codium, *Stackh.*
283. bursa, *A.*
284. tomentosum, *Stackh.*
285. amphibium, *Moore.*
286. adhærens, *A.*

CV. Vaucheria, *G.*
287. submarina, *B.*
288. marina, *Lb.*

CVI. Bryopsis, *G.*
289. plumosa, *A.*
290. hypnoides, *La.*

XXII. Ulvaceæ.

CVII. Porphyra, A.
291. vulgaris, A.

CVIII. Bangia, Lb.
292. fusco-purpurea, Lb.
293. ciliaris, Carm.
294. ceramicola, Chauvin.
295. elegans, Chauvin.

CIX. Enteromorpha, Link.
296. cornucopiæ, Hooker.
297. intestinalis, Link.
298. compressa, G.
299. clathrata, G.
 v. Linkiana, J. E. G.
 (E. Linkiana, P. B.)
 erecta, J. E. G.
 (E. erecta, P. B.)
 ramulosa, J. E. G.
 (E. ramulosa, P. B.)
300. Hopkirkii, M'Calla.
301. percursa, Hooker.
302. Ralfsii, H.

CX. Ulva, L.
303. linza, L.
304. latissima, L.
305. lactuca, L.

XXIII. Confervaceæ.

CXI. Leptocystea, J. E. G.
306. pellucida, J. E. G.

CXII. Cladophora, K.
307. rupestris, K.
 v. β. distorta.
308. rectangularis, Griff.
309. lætevirens, K.
310. diffusa, H.
311. Hutchinsiæ, H.
312. Macallana, H.
313. falcata, H.
314. glaucescens, Griff.
315. flexuosa, Griff.
316. refracta, K.

317. albida, K.
318. gracilis, Griff.
319. Brownii, H.
320. Magdalenæ, H.
321. flavescens, K.
322. nuda, H.
323. Balliana, H.
324. fracta, K.
325. arcta, K.
326. lanosa, K.
327. uncialis, H.
328. Gattyæ, H.
329. repens, J. A.
330. Rudolphiana, K.

CXIII. Chætomorpha, K.
 (Conferva, P. B.)
331. melagonium, K.
332. ærea, K.
333. sutoria, B.
334. linum, K.
335. tortuosa, K.
336. implexa, K.
337. arenicola, B.
338. arenosa, Carm.

CXIV. Cytophora, J. E. G.
 (Conferva, P. B.)
339. litorea, J. E. G.

CXV. Hormotrichum, K.
340. Younganum, K.
 (Conferva Y., P. B.)
341. collabens, K.
 (Conferva c., P. B.)
342. bangioides, K.
 (Conferva b., P. B.)
343. Cutleriæ, H.
 (Lyngbya C., P. B.)
344. speciosum, J. E. G.
 (Lyngbya s., P. B.)
345. Carmichaelii, K.
 (Lyngbya C., P. B.)

CXVI. Rhizoclonium, K.
346. riparium, K.
347. Casparyi, H.

348. flaccum, J. E. G.
 (Lyngbya f., P. B.)

XXIV. Oscillatorieæ.

CXVII. Lyngbya, A.
349. majuscula, H.
350. ferruginea, A.

CXVIII. Calothrix, A.
351. confervicola, A.
352. luteola, G.
353. scopulorum, A.
354. pannosa, A.
355. semiplena, A.

CXIX. Oscillatoria, Vaucher.
356. litoralis, Carm.
357. spiralis, Carm.
358. nigro-viridis, Thwaites.
359. subuliformis, Thwaites.
360. insignis, Thwaites.

CXX. Spirulina, Turpin.
361. tenuissima, K.

CXXI. Tolypothrix, K.
362. fasciculata, J. E. G.
 (Calothrix f., P. B.)

CXXII. Arthronema, Hassall.
363. hypnoides, J. E. G.
 (Calothrix h., P. B.)
364. cæspitula, J. E. G.
 (Calothrix c., P. B.)

CXXIII. Microcoleus, Desmaz.
365. anguiformis, H.

CXXIV. Schizothrix, K.
366. Cresswellii, H.

LIST OF ORDERS, GENERA, AND SPECIES. xxiii

CXXV. Schizosiphon, K.
367. Warreniæ, *Caspary.*
CXXVI. Rivularia, *Roth.*
368. plicata, *Carm.*
369. atra, *Roth.*
370. nitida, *A.*
CXXVII. Actinothrix, *J. E. G.*
371. Stokesiana, *J. E. G.*

XXV. **Nostochineæ.**
CXXVIII. Monormia, *B.*
372. intricata, *B.*
CXXIX. Sphærozyga, *A.*
373. Carmichaelii, *H.*
374. Thwaitesii, *H.*
375. Broomei, *Thwaites.*
376. Berkeleyana, *Thwaites.*

CXXX. Spermosira, *K.*
377. litorea, *K.*
378. Harveyana, *Thwaites.*

XXVI. **Bulbochæ-taceæ.**
CXXXI. Ochlochæte, *Thwaites.*
379. hystrix, *Thwaites.*

ERRATA.

Page 106, line 22, *for* Richardsonii *read* Richardsoni.
  " 173, line 26, and page 174, lines 3 and 12, *for* Stenogramme *read* Stenogramma.
  " 214, line 19, *for* XCXIX. *read* XCIX.

BRITISH SEA-WEEDS.

CHAPTER I.

THE POSITION OF SEA-WEEDS IN THE VEGETABLE KINGDOM; THEIR STRUCTURE, ETC.

A PLANT, according to recent authors, is "a cellular body, possessing vitality, living by absorption through its outer surface, and secreting starch." Accepting this definition as sufficient for my present purpose, which is not to treat of plants generally, but only of a portion of a single class, I will trace a brief outline of the natural arrangement, and try to mark distinctly the position in which Sea-weeds stand in relation to other plants, whether of higher or lower organization.

The comprehensive definition which I have quoted is purposely framed to include the whole of what is aptly called the Vegetable Kingdom, from the lordly Oak and mighty Wellingtonia, which take a century to reach their full growth, to the insignificant Algæ which spring into life, flourish, propagate, and decay in the course of a few days on the surface of some chance puddle of stagnant water.

This Vegetable Kingdom, in analogy with its prototype, is divided, not into provinces, counties, parishes, etc., but into classes, orders, genera, and species, based on and in accordance with the all-wise unchanging natural laws by which it is governed.

The first division is into the following classes:—

1. *Exogens*, or *Dicotyledons*, whose stems increase in thickness by the formation of concentric layers of new wood between the old and the bark; whose leaves have branched veins; and whose seeds have two separate germs—one for the stem, the other for the root.

This class includes most of the trees, bushes, and herbaceous plants of temperate climates.

2. *Endogens*, or *Monocotyledons*, whose stems increase longitudinally, without becoming much thicker; whose leaves have parallel veins; and whose seeds have but one germ. Palms, Aloes, Grasses, and most bulbous-rooted plants belong to this class.

3. *Acrogens*, or *Cryptogamic Plants*, which differ from both the other classes in almost every particular. They have neither stems, leaves, flowers, nor seeds, properly so called, but have instead differently constituted organs, answering the same purposes.

All flowerless plants, from the large Tree-ferns of the tropics to the minute one-celled Algæ of our ponds and ditches, are included in this class; and there is in consequence much greater difference between the organization of its highest and lowest forms than is to be found in either of the preceding classes.

Dr. Lindley, in his 'Introduction to Botany,' divides *Acrogens* into six Alliances, which he calls respectively, Filicals (Ferns, etc.); Lycopods (Club-mosses and Pepperworts); Muscals (Mosses, etc.); Lichens; Fungals

(Fungi); and Algals (Confervæ, Seaweeds, etc.); so that Sea-weeds are nearly, not quite, the lowest forms of vegetable organization. Numerous freshwater Algæ and the large families of *Diatomaceæ* and *Desmidiaceæ* are lower still, and for the present at least must be considered to be at the base of the scale.

Having thus traced the humble relationship of Seaweeds to their kindred plants, I proceed to describe the peculiarities on which that relationship is based. In doing so I shall draw largely on Dr. Harvey's admirable introduction to his 'Nereis Boreali-Americana,' and I take this opportunity to record my deep sense of gratitude for the information that I have derived from that valuable Essay on this interesting but difficult subject.

All plants are composed of cells variously developed, according to the functions they have to perform in the formation of roots, wood, bark, stems, leaves, or flowers.

In Flowering-plants and Ferns, *vascular* plants as they are called in consequence of the presence of *vascular* tissue in their structure, the cells assume many varieties of form and substance; but in *cellular* plants, Mosses, Lichens, Fungi, and Algæ, they are more uniform and simple, and in Algæ, including of course Seaweeds, they are all of the same nature as those found in the soft parts, the leaves for instance of plants of the higher orders.

In the Vegetable Kingdom there are however no hard lines of demarcation; the descent from the highest to the lowest form is gradual, and the arrangement of the cells in the more perfect Sea-weeds, though different from, is still analogous with that observed in *vascular* plants. Organs similar to roots, stems, and leaves are still present, and are furnished with different series of cells, answering to the wood, bark, veins, etc., of the stems and leaves of trees and plants. " Passing from such," writes Dr. Harvey, " we meet with others gradually less and less perfect, until the whole vegetable is reduced either to a root-like body, or a branching, naked stem, or an expanded leaf; as if Nature had first formed the types of the compound vegetable organs so named, and exhibited them as separate vegetables; and then, by combining them in a single framework, had built up her perfect idea of a fully-organized plant. But among Algæ we may go still lower in vegetable organization, and arrive at plants where the whole body is composed of a few cells strung together, and finally at others— the simplest of known vegetables—whose whole framework is a single *cell.*"

The *root* of a Sea-weed is either in the form of a disc, or, more rarely, fibrous. It is small in proportion to the size of the plant, and in some species is quite obscure, or altogether absent.

It does not absorb and transmit nourishment, like the root of a more highly-organized plant, and indeed its sole function is to serve as a means of attachment. It does not penetrate the rock or other substance to which it may be affixed, but simply clings to the surface. This it does more or less firmly in proportion to the resistance that it has to offer to the action of the ever-moving sea, and great indeed must be the power which the roots of the larger *Fuci* have to exert in order to maintain their position on rocks exposed to the full sweep of the ocean waves in all their vicissitudes of mood, from calm to raging storm.

The term *frond*, strictly speaking, includes all the parts of a Sea-weed, except the *root* and the *fructification.*

INTRODUCTION.

The *fronds* of different genera vary much in shape and character. Some consist wholly of a flat, leaf-like membrane, as *Ulva* and *Porphyra;* others are simple membranous sacs, as *Enteromorpha* and *Asperococcus;* some are cylindrical and solid throughout, while others are so only in the lower part, or stem, and have expanded leafy processes above. A few are similar in appearance to lichens and fungi, while others, again, are scarcely to be distinguished from the rocks on which they grow.

In the higher or compound forms the *frond* is generally plant-like, and furnished with stem, branches, and branchlets. The stem is in most cases composed of two or more series of cells, and if these cells are all of the same length, the stem has the appearance of being jointed or articulated. If the cells are of unequal length, no joints are apparent, and the stem is said to be inarticulate. The allied genera *Rhodomela* and *Polysiphonia* are examples of these two kinds of *fronds,* those of the former being inarticulate, those of the latter articulate. This cellular arrangement of the interior of the *frond* is a very important element in the study of Sea-weeds, and is a character much relied on in the determination of orders, genera, and even species.

In order to ascertain the nature and position of the cells, it is necessary to examine thin transverse and longitudinal sections of the different parts of the *frond* under a microscope. In some genera they are uniform in size, and all arranged longitudinally; but more commonly they are divided into two principal series, that in the centre called the *axial* series, that on the surface the *peripheric* or *cortical* series. The former series is generally composed of elongate thread-like cells, arranged lengthwise, either packed closely together or separated by intervening layers of gelatine, and is spoken of as the *axis* of the *frond*. The cells of the *peripheric* series are for the most part smaller, and are disposed horizontally. Both these series are subject to numerous modifications in different genera, and in many instances subsidiary series are interposed.

The gelatine, which enters largely into the composition of the *fronds* of Sea-weeds, varies very much, both in quantity and quality, in different genera. In some it is almost as limpid as water, and in others nearly as firm as cartilage. In some, as for instance *Dudresnaia*, it so overpowers the other constituents of the *frond* that the whole plant seems to be but a mass of slime, while in contrast to this there are many genera where it is scarcely perceptible.

It is this gelatine which gives substance to the *frond*, and hence those plants, where the quantity is small, are said to be *membranaceous*; those where it is abundant and fluid, *gelatinous*; and those where it is firm, *cartilaginous*. The shrinking which all Sea-weeds undergo in the process of drying is due to the expulsion of this element of their composition.

The mode of growth is not the same in all Sea-weeds. The *fronds* of some genera grow only at the tip, and their youngest shoots are, therefore, those which are furthest from the root. In the *Laminariæ*, on the other hand, the new growth takes place at the base of the old *frond*, and for this reason it is almost impossible to find any but very young specimens with perfect tips. In several species of this genus a new *frond* is formed, and the old one thrown off, every season; and specimens of *L. digitata*, showing this process in progress, are not uncommon, and are well worthy of a place in every collection.

The *fronds* of some of the filiform genera are deciduous, and their branchlets fall off like leaves in Autumn, causing the same kind of difference between the summer and winter aspect of the plant as there is between a tree covered with leaves, and one whose branches are bare. Several species of *Rhodomela* and *Polysiphonia* have *fronds* of this description.

There are other genera, the young *fronds* of which are clothed with tufts of delicate, jointed fibres, that fall off as the *frond* becomes mature. These fibres, which are most strikingly obvious in *Desmarestia*, *Sporochnus*, and *Arthrocladia*, but exist in variously modified forms in genera widely separated by other characters, are almost always found on the growing part of the *frond*, and it is therefore supposed that their functions are similar to those of the leaves of higher plants.

One other organ that I must notice is the *air-vessel*, or *float*, which occurs almost exclusively among plants of the olive series. It assumes various forms, but is always adapted to increase the buoyancy of the stem and branches, and so to lessen the resistance which they offer to the advancing or receding waves. In some genera these vessels are simple cavities, formed by swellings of the stem or branch, and filled with air, as in the common Popweed (*Fucus nodosus*). In others, several placed near together merge into a single, chambered vessel, as in *Halidrys siliquosa*. In some of the *Sargassa* they are borne, berry-like, at the tips of short branches. This is notably the case with the Gulf-weed, which derives its specific name of *bacciferum*, or *berry-bearing*, from these organs. In addition to those forms of *air-vessels*, which are constant to the particular genera and species to which they belong, there are others which are occasionally developed in an abnormal manner, to meet, as it were, the exceptional requirement of some plant that does not usually bear them, as, for example, at the tips of the *fronds* of large, deep-water specimens of *Chorda filum*.

I have endeavoured to point out the most important peculiarities in the form and structure of the *frond*, as it is developed in our "native seaweeds," but, although these afford a very fair epitome of the whole marine flora, it must be borne in mind that many examples of brilliant colouring, magnificent size, and exquisite delicacy, both of texture and mechanism, are to be found in exotic genera, no representatives of which exist on our shores.

The *fructification* of Sea-weeds is closely connected with, and to a certain extent dependent on, the structure of the frond, and the consideration of it is of the highest importance in the pursuit of an accurate systematic knowledge of the class.

Speaking in general terms of the process of fruiting among the higher *Algæ*, Dr. Harvey writes as follows:—
"*Spores*, or *sporangia*, appear to be formed by certain cells attracting to themselves the contents of adjacent cells; and, in the compound kinds, empty cells are almost always found in the neighbourhood of the fruit-cells; but, with the complication of the parts of the frond, the exact mode in which spores are formed becomes more difficult of observation. At length, among the highest Algæ, we encounter what appear to be really two sexes, one analogous to the *anther*, and the other to the *pistil* of flowering plants. It would seem,

however, that it is not each individual spore which is fertilized, as is the case in seed-bearing plants; but that the fertilizing influence is imparted to the pistil, or sporangium itself, when that body is in its most elementary form, long before any spore is produced in its substance, and even when it is itself scarcely to be distinguished from an ordinary cell. *Antheridia*, as the supposed fertilizing organs are called, are most readily seen among the *Fucaceæ*."

After describing a second mode found only among the simpler kinds, where the whole body consists of a single cell, Dr. Harvey proceeds:—"The third mode of continuing the species has been observed in many Algæ of the *Green* series, in some of which sporangia are also formed, but in others no fructification other than what I am about to describe has been detected. This mode is as follows. In an early stage the green matter, or *endochrome*, contained within the cells of these Algæ, is of a nearly homogeneous consistence throughout, and nearly fluid; but at an advanced period it becomes more and more granulated. The granules, when formed in the cells, at first adhere to the inner surface of the membranous wall, but soon detach themselves, and float freely in the cell. At first they are of irregular shapes, but they gradually become spheroidal. They then congregate into a dense mass in the centre of the cell, and a movement, aptly compared to that of the swarming of bees round their queen, begins to take place. One by one these active granules detach themselves from the swarm, and move about in the vacant space of the cell with great vivacity. Continually pushing against the sides of the cell-wall, they at length pierce it, and issue from their prison into the surrounding fluid, where their seemingly spontaneous movements are continued for some time. The vivacious granules, or *zoospores*, as they have been called, at length become fixed to some submerged object, where they soon begin to develope cells, and at length grow into Algæ similar to those from whose cells they issued."

The organs whose production Dr. Harvey has thus lucidly described are analogous to the seeds of land plants; but it is evident that, although they perform the same function, they are altogether of a different nature. In the descriptive part of this work they will be spoken of as *spores, tetraspores, antheridia,* and *zoospores,* and a great many important characters will be found to depend on them.

A vast amount of study has been devoted to the elucidation of the early and obscure stages of the development of Sea-weeds; but, owing to the difficulty of conducting observations, the results obtained are not so definite as those ordinarily derived from scientific research. Various writers do not agree as to the exact nature of the different kinds of fruit, and the precise part they play in Nature's drama.

In the *Olive* series (*Melanospermeæ*), the *fructification* consists wholly of *spores* and *antheridia*, which are produced either on the same or on separate plants. If the former, it is said to be *monœcious;* if the latter, *diœcious.* The *spores* are olive-coloured, and are formed from delicate, jointed fibres (*paranemata*). They are enveloped in a transparent membrane, called a *perispore.* Each *perispore* contains what, in its early state, appears to be a single spore; but, as this matures, it either retains its individuality, or separates into two, four, or eight *sporules.* The *spores,* with their *peri-*

spores, are either enclosed in spore cavities (*conceptacles*), or arranged externally, singly or in groups (*sori*). The details of this part of the subject will be given in the description of the Orders and genera.

The origin of the *antheridia* is very similar to that of the *spores*. Both spring from and among the *paranemata*, and sometimes grow side by side in the same conceptacle. The subsequent development of the *antheridium* is, however, different. In the place of the *spore* and *perispore*, a large transparent closed cell, filled with a number of very minute bodies (*sporidia*) is formed. These *sporidia* are analogous to the pollen of flowering plants; but it is not yet decided whether they perform the same fertilizing functions. Some writers, the majority, I think, are of opinion that they do, whilst others maintain that they are only gemmules, and themselves capable of germination. Without attempting to decide where doctors differ, I will again transcribe from Professor Harvey the following interesting account of them:—"They (the *antheridia*) are oval, somewhat pointed at one end, and contain a reddish-orange granule; and they are furnished with two extremely slender vibratile hairs, or cilia, one of which issues from the narrow extremity of the corpuscle; the other, which is of greater length, from the coloured granule. The corpuscles, at first contained within the *antheridium*, at length issue from it, escaping into the surrounding water, and immediately commence a succession of rapid movements to and fro, and in circles and curved lines, strikingly similar to the ciliary movements of some of the *Infusoria*, or of the *spores* (*zoospores*) of some of the freshwater Algæ of the *Green* series. These movements depend on the rapid vibrations of the cilia. During progression, the narrow end of the corpuscle is always in front; while the cilium rising from the coloured granule trails behind, like a tail."

In the *Red* series (*Rhodospermeæ*) *spores, tetraspores,* and *antheridia* are present, the two former in all, the latter only in a part of the species. The development of two kinds of fertile fruit on different individuals of the same species is the chief character of the series, and that it should prove to be exactly coincident in general range with that of colour which had been previously adopted is a striking proof that both systems of classification are correct. It would appear that each species of this series is composed of two distinct parts, each perfect in itself, and capable of independent reproduction, yet both so evidently identical as to form inseparable parts of the same whole. *Spores* and *tetraspores* are not "different phases of the same organ, but are in their origin and development perfectly distinct, and formed with the greatest regularity, following fixed laws," neither are they equivalent to the stamens and pistils of diœcious plants of higher Orders. They may rather be considered, the one a true *spore*, supposed to be fertilized by means of an *antheridium;* the other, a mere *gemmule,* or bud of the simplest possible structure, which is cast off by the parent plant, and carries with it sufficient vitality to become the nucleus of a fresh individual. It is far from certain that this is a true account of the nature of these bodies; and even if it be, there is the further question to be answered, which is the *spore* and which the *gemmule?* On this latter point there is a wide divergence of opinion, and as the evidence on both sides is pretty evenly balanced, it is not wonderful that the deductions from it should vary.

The *spores* are always arranged in a more or less definite mass, or tuft, called the *spore-cluster*, or *sporiferous nucleus*. Each *spore* is formed of a single cellule, with a transparent membrane, and containing a nearly solid mass of a dark-coloured starch-like substance (*endochrome*), which, on being expressed, breaks up into impalpable granular dust. There are two kinds of *spore-clusters*. In that which is found in the most perfect Orders of the series, the nucleus consists of a tuft of jointed necklace-like fibres (*spore-threads*), which radiate in all directions from a central point, or grow from a *placenta* lodged in a *conceptacle*. The *spores* are developed in the cells of the *spore-threads*, one or several on the same thread, but only one in each cell. "In the less organized families," writes Dr. Harvey, "the nucleus is formed either from a single mother-cell, from several detached mother-cells, or from such cells imperfectly joined together in moniliform strings issuing from a central joint, or growing from the placenta of a conceptacle. Each mother-cell, which is at first filled with a homogeneous endochrome, becomes by repeated cell-division converted into a cluster of *spores*, at first retained within its walls; afterwards, on the bursting of the wall, dispersed. Thus, by the evolution of one cell, a *favella*, or simple globose nucleus, containing many angular spores within a hyaline periderm, is formed. By the evolution of several detached, but adjacent, mother-cells, a *favellidium*, or compound *favella*, results; and by the similar evolution of the cells of the moniliform series, the highest form of *favellidium* is produced. In all these cases the general *nucleus*, as well as the particular *nucleoli*, is surrounded by a gelatinous or submembranaceous hyaline periderm, derived from the cell-walls of the transformed cells."

As the present arrangement of the Red weeds is founded on the structure of the *spore-clusters*, it is very important that the student should become thoroughly acquainted with these organs. In the various species of such genera as *Polysiphonia*, *Laurencia*, etc., they may be easily examined under a microscope of moderate power, by simply pressing the conceptacle between two slips of glass, and in most of the more perfect divisions of the series they may be exhibited either in longitudinal or transverse sections of the conceptacles. Among the species of the less perfect section the process of examination is more difficult, but a little practice and perseverance will generally yield sufficient results to answer the purpose of all ordinary collectors.

The *tetraspores* are formed either from the bark-cells or from the upper branchlets. They are very variously arranged in different genera. In some they are scattered singly among the surface-cells over the whole frond, in others they are collected in more or less distinct clusters (*sori*); in others, again, they are found only in the branchlets; or in external warts (*nemathecia*); or in special leaflets (*sporophylla*); or in pod-like receptacles (*stichidia*). They are always produced in a constant regular manner, in accordance with fixed natural laws, and are all constructed on the same principle. Each consists of a dark-coloured mass of *endochrome*, enveloped in a transparent, membranous sac, and marked by lines for division into four parts. This mass, when mature, separates into four, or rarely into eight or more *sporules*. The divisions into four—whence the name of *tetraspore*, which is derived from the Greek *tetras*, *four*, and *sporos* a *seed*—are variously made in different genera, and are frequently used as distinctive characters. There are three principal modes. In the first, the *spore* is simply quartered by two transverse lines, crossing at right angles at its centre, and is said to be *cruciate*. In the second, it is divided into four unequal parts, three of which only are visible at once, by lines which radiate from the centre, and is called *tripartite*. The third and last kind is called *zonate*, and is transversely divided.

The annexed woodcuts of the appearance of the different kinds, when highly magnified, will be more easily understood than my imperfect description.

Fig. 1. Cruciate. Fig. 2. Tripartite. Fig. 3. Zonate.

The *fructification* of the Green series is wrapt in greater mystery than even that of the Reds. It consists chiefly of *zoospores*, which have been already described; but ordinary *spores* and *antheridia* are also developed in certain genera. The existing knowledge of the subject is too imperfect to afford the data for a popular detailed description; and, further, it is almost impossible to separate that part which relates to the marine species from that belonging only to the fresh-water.

I trust that I have traced a sufficiently clear outline of the mode of *fructification* to serve the purpose of the majority of my readers. Those who desire to extend their researches I must refer to the works of Agardh, Harvey, and other writers.

CHAPTER II.

ON THE COLOUR, DISTRIBUTION, ETC., OF SEA-WEEDS.

AMONG the higher orders of plants, the different parts of the same individual vary in colour; but the same part in different individuals is always approximately of the same hue. Thus the trunk, or stem, is generally of some shade of brown, the leaf green, and the flower of a brighter colour.

Among Sea-weeds, on the other hand, the colour of all the parts of an individual is the same, while that of individuals—of different families, of course—is different.

This variation has been used as a basis of classification, and all Sea-weeds are divided into Olive-green weeds, or *Melanospermeæ*; Red weeds, or *Rhodospermeæ*; and Grass-green weeds, or *Chlorospermeæ*. Several purple weeds are included in the last class; and beside this there are others which are not in all circumstances constant to the colour of the class to which they belong.

The *Olive-green weeds* are for the most part of large size, and very abundant. They grow chiefly between, and a little beyond, the tide-marks, and furnish the great bulk of our shore vegetation. Some exotic species grow at a great depth, and are of enormous size.

The *Red Weeds* are, as a rule, of delicate structure, and grow in deep water, or in pools, or other sheltered positions. Their colour is fugacious, and is frequently destroyed by exposure to the influence of light. The effect produced on them by this means differs in degree according to the extent and duration of the exposure. Some fronds are only reduced a few shades paler than their natural tint; others are turned yellow or green; and others are thoroughly bleached; but this extreme is not usually reached except in the case of dead specimens.

It is a common error among inexperienced collectors to believe that plants in this abnormal—I might almost say diseased—state are generically or specifically different from their bright-coloured, healthy brethren; but this is a delusion which a little practice will dispel.

The class of *Grass-green Weeds* includes the remainder of the Sea-weeds, and in addition a very large number of fresh-water species. The marine portion, with which alone I am at present concerned, grow in shallow pools, and on the sides of rocks near the surface, but not usually where they are altogether deprived of water by the receding tide. Many species, notably some of the *Enteromorphæ*, seem to grow indifferently in quite fresh or quite salt, or any intermediate degree of brackish water, and nearly all the class love the light, and rather gain than lose colour when exposed to the brightest rays of the sun. There are, however, deep-water species which boast as bright a green as their companions on the shore, and it is therefore evident that light is not an indispensable agent in the production of their colour.

Several species of *Olive* and *Red* weeds are very beautifully prismatic when in the water, as the light falls on

c

them in different directions, and exhibit a succession of the brightest metallic tints of green and blue and purple. *Cystoseira ericoides* is the most remarkable example of this effect, which is one of its specific characters. The phenomenon ceases directly the plant is drawn from the water, and the collector is aware that it must do so. Nevertheless, he will probably experience a momentary disappointment to find his dusky beauty so suddenly despoiled of all her brilliant gems.

There are various processes, such as the application of heat, steeping in fresh water, etc., by which the natural colour of Sea-weeds may be heightened or changed; but I only mention them in order to caution my readers that they may be applicable to the production of artistic effects in Sea-weed pictures, baskets, etc., but should never be used in the preparation of scientific specimens.

I have alluded to the general distribution of the three classes of Sea-weeds—the *Olive-greens* on the shore and exposed rocks; the *Reds* in deep water or shady pools; and the *Grass-greens* near high-tide mark, where they are subject to the action of light, and to contact with fresh water. These are facts which should interest every student of marine botany, whether his collection and researches be general or confined to the plants of any particular locality. The Sea-weeds of our shores are but a part of the large family distributed over all the waters of the globe, of which they are, as it were, an epitome. There is a natural harmony running through the whole, and the same laws, modified, of course, by circumstances, apply to the waters which surround these islands, as to those which beat upon the shores of the Antipodes.

There are certain physical influences which affect Sea-weeds, and to some of these I will now refer as briefly as possible.

I will first revert for a moment to the whole family of *Algæ*, to state that the only conditions indispensable to their existence are the presence of air and moisture. They will grow in the waters of mineral springs, no matter how hot or how cold; in vats of poisonous chemical solutions; in dark caves; and amid the Arctic snows. Even the nebulous vapours high above the surface of the earth are supposed to afford them a sufficiently hospitable shelter. Notwithstanding, however, this almost universal diffusion of certain of the lower Orders, the more perfect members of the class are all natives of the sea, and thence derive their name of *Algæ*, or *Sea-weeds*, the latter name being but a translation of the former. With few and unimportant exceptions, such as *Zostera marina*,—the grass which grows so abundantly on many muddy shores between the tide-marks, and which was proposed as a substitute for cotton during the recent scarcity of that fibre,—the entire vegetation of the ocean is composed of Sea-weeds. The hardy *Fuci* make their appearance on the first rocks that escape the rigorous grasp of perpetual ice in Arctic and Antarctic latitudes, and thence with ever-increasing variety of forms, and, as a rule, with greater and greater luxuriance of growth, the Nereid multitude rolls on north or south until it culminates in the perfection of its fairy radiance and beauty in the hottest regions of the tropical seas. Rocks are everywhere the chosen haunts of the majority of Sea-weeds, which do not appear to have much predilection for any particular kind; nor indeed is this wonderful, for they seek them for shelter and for foothold, not for nourishment. But though the material of the rock is unimportant, the conformation is not; and those which are rough and rugged, and abound in tiny bays and creeks, are always more productive than those with smooth surfaces, perpendicular sides, and regular outline. Not only rocks, but mud, sand, shells, floating or submerged wood, the copper sheathing or the iron sides of ships all have their tenants; nor do these suffice, for even yet some of our Nereids are homeless, and the mighty mass of *Sargassa*, known as the Gulf-weed, grows unattached floating on the surface of the sea, and there, unaided by roots or, I believe, by *spores* or *tetraspores*, has attained a bulk which equals, if, indeed, it does not exceed, that of any other single kind throughout the world.

The depth to which Sea-weeds extend cannot be very exactly defined; but there is no doubt that, as a rule, they only form a fringe a mile or so wide around the land. In cold and temperate climates a depth of a dozen fathoms would be probably a liberal allowance for any of the more perfect forms, but in tropical and sub-tropical latitudes perhaps twice or thrice that depth would not be an exaggeration, and in exceptional instances even these limits are occasionally exceeded.

These remarks do not apply to the lower forms of *Algæ*, for myriads of *Diatomaceæ* have been found to exist at every depth of the ocean which has hitherto been fathomed. There are also a vast number of calcareous plants, belonging to the family of *Corallines*, whose substance is principally composed of lime extracted from the sea-water. These grow at considerably greater depths than sea-weeds proper, and are, with the exception of *Diatomaceæ*, the sole representatives of the Vegetable Kingdom, over considerable regions of the bottom of the sea.

The form and character of a particular species are frequently much changed in specimens grown either at a greater or less depth than usual. In the majority of instances there would be a tendency to increase in luxuriance *pari passu* with the increase of depth. A plant whose ordinary habitat was near high-water mark, would attain a larger size in deep water, and one from beyond the limit of the tide would be dwarfed by transfer to a shallow pool. There are, however, instances where the change would be in an opposite direction, and the denizen of the shore would become stunted in deep water.

The bed of the ocean resembles in a great degree the surface of the earth. It has its mountains and its valleys, its plains and deserts, its various kinds of rocks and soils; and these physical peculiarities affect vegetation in the sea very much in the same manner as they do on land. Submarine mountains, valleys, or deserts hinder the diffusion of the *spores* of Sea-weeds from one coast to another, just as those on land interrupt the spread of seeds. On the other hand, *spores* have the water as a means of transit, and *seeds* the air; but while the former can all float, only a portion of the latter are furnished with the means of flight. Hence the barrier opposed to the migration of Sea-weeds is less complete than it is to that of terrestrial plants.

The temperature of the sea is less liable to variation than that of the air, and the effect of climate on Sea-weeds is not in consequence great. Some Orders and genera are more or less confined to cold or temperate, and some to warm regions, while others are more generally diffused; the different species of the same genus, and occasionally even the same species, extending over a wide range of latitude. Nature, it would seem, does not love hard boundary-lines, and among Sea-weeds, as in her other works, one form fades into another by slow and almost imperceptible degrees. The marine flora of our own shores offers a happy illustration of this law, for while it embraces all the hardy, cold-loving species, it also includes a very large proportion of delicate southern forms, which reach their northern limit among the sheltered bays of Devon, Cornwall, the South and West of Ireland, and the Channel Islands.

Oceanic currents have a very powerful influence on the distribution of Sea-weeds. They carry masses of water above or below the mean temperature far beyond the limits where it would otherwise exist, and thus extend the range of habitats suitable for particular genera and species. They also convey the plants themselves, or their spores, to new localities along the shores round which they flow, and even from one coast to another, when they cross a channel, a gulf, or any larger tract of sea. The great Gulf Stream is the most notable example of this agency, and its effect is very visible around these Islands. The coasts of Devon and Cornwall and the Western shores of Ireland are, with the exception of the Channel Islands, the chief seats of our Algological wealth; and it is a curious fact that the same tender species which appear to maintain with difficulty a precarious footing on the extreme southern corner of England, grow more boldly in Ireland in a considerably higher latitude. The explanation of this apparent inconsistency is doubtless to be found in the presence of the warmer waters of the Gulf Stream, which flows in that direction.

I have hitherto spoken of the effects of average climate or temperature. I must not omit to mention that of exceptional intensity either of heat or cold. A long cold winter or a brilliant burning summer do not act so directly nor with such force on Sea-weeds as they do on gardens and cornfields. Nevertheless the one will sometimes hinder the development of those more southern species which ordinarily attain a modified maturity on our shores; and the other, on the contrary, will force them into the full perfection which characterizes them in their more genial home.

> " The penetrative Sun,
> His force deep darting to the dark retreat
> Of vegetation, sets the steaming power
> At large."—*Thomson.*

CHAPTER III.

ON COLLECTING AND PRESERVING SEA-WEEDS, THEIR USES, ETC.

> "As we strolled along,
> It was our occupation to observe
> Such objects as the waves had tossed ashore,
> Tangle, or weed of various hues and forms,
> Each on the other heaped, along the line
> Of the dry wrack. And, in our vacant mood,
> Not seldom did we stop at some clear pool
> Hewn in the rock, and, wrapt in pleasing trance,
> Survey the novel forms that hung its sides,
> Or floated on its surface,—too fair
> Either to be divided from the place
> On which they grew, or to be left alone
> To their own beauty."

It is not possible to prescribe precise rules which shall be applicable to the collection of all kinds of Sea-weeds in every varied circumstance of coast and season, or to the opportunities or object of each individual collector. Experience will be found to be the best teacher of all minor details, and the time and labour expended to obtain it will be productive of many collateral advantages; one of the chief of which will be a practical knowledge of the habit and appearance of various species, which can only be obtained by observing a large number of growing specimens.

A few general observations, the result of a long apprenticeship, may, however, be useful. The dress of a zealous collector should consist of as few and as coarse, unspoilable garments as possible, for he will most probably get more or less wet, even if he be not tempted to wade; he should be furnished with one or two small tin botanical boxes, a couple of broad-mouthed, two-ounce, boxwood-topped bottles, an oyster-knife, a good stout walking-stick, a pocket lens, and a penknife.

Where there is an option, it is well to start for the shore so as to arrive there about an hour before low water, and to follow the sea as it recedes. All the precious moments of the extreme ebb of the tide should be devoted to searching the furthest rocks that can be reached, as it is on these that the most delicate rare weeds, particularly of the Red series, grow. The more patient the search, the greater will be the reward. Every pool, creek, cave, and overhanging or perpendicular rock, should be carefully examined; all large, coarse weeds should be pushed aside, or removed, so as to reveal any of the smaller species that may be sheltered beneath them, or may grow parasitically on their roots, stems, and fronds. It is important, when circumstances permit, to obtain the whole plant with the root attached, and, in the case of the larger species, which it is impossible to preserve entire in any ordinary herbarium, the specimens should be so arranged as to include all the parts, from the root upwards if possible, and to exhibit the character and mode of growth,—the mere fragments of the upper branches, which are sometimes collected, being comparatively worthless for all scientific purposes. In order to avoid carrying home useless matter, which is very detrimental to the examination or successful preservation of specimens, each plant, as it is gathered, should be washed in some shore pool, to free it from sand or other extraneous substance, and the redundant or broken branches should be removed. It should then be placed in the box, in such a manner that it will remain separate and not be damaged by or become entangled with its fellows. Specimens of *Desmarestia, Sporochnus,* and *Arthrocladia* must be kept by themselves, for they not only decompose very rapidly, but also cause other weeds with which they are in contact to do the same. *Polysiphonias, Dasyas,* and *Griffithsias,* indeed most of the more delicate Red weeds, lose their colour if put into fresh-water. They should therefore be kept separate, and I have always found it convenient to carry them in a bottle filled with sea-water.

There are three modes in which specimens may be obtained,—first and best by gathering them in a growing state; secondly, by picking up such as are cast on shore by the sea; and, lastly, by dredging. The two former are more accessible than the latter, and a very large collection may be formed by following them. There are, however, a few deep-water species which can rarely be obtained, except by means of the dredge. During spring-tides, which occur at new and full moon, many rocks and pools which are ordinarily out of reach may be explored, and the opportunities thus afforded should always be diligently used. Again, after a gale, particularly if it should be coincident with the spring-tides, many rare deep-water species are thrown up, and may be found among the masses of weed left by the receding waves.

> " When descends on the Atlantic
>   The gigantic
>   Storm-wind of the equinox,
>   Landward in his wrath he scourges
>   The toiling surges,
>   Laden with sea-weed from the rocks."—*Longfellow.*

The season best adapted for collecting Sea-weeds is from the beginning of May to the end of September; but there is no time of the year which may not be profitably employed on the shore. There are many kinds which are only to be found in their most perfect state in winter, and there are others which then assume a totally different aspect to that they wear in summer. As with the season, so also with the locality. There are certain districts, particularly in the south and west of England and Ireland, and all round the Channel Islands, which are more highly favoured than others; but there are plenty of good weeds to be found on nearly every part of our coast, and places which to the eye of the young collector are least promising, are often prolific of interesting species, and afford the best opportunity for observing the varying appearances and modes of growth of different genera.

If, for instance, he should be at West Cowes in early spring, and will look over the sea-wall of the parade, he will find it fringed with dwarf plants of *Fucus vesiculosus* and the boulders beneath covered with larger specimens of the same species, or with the wig-like narrow-leafed *Fucus nodosus,* on which latter the pretty little *Polysiphonia fastigiata* will appear in dark brown tufts. If he will descend and examine the wall and boulders more closely, searching in the less exposed parts, and lifting or cutting away the *Fuci,* so as to lay bare what may be underneath, he will find a fresh series of interesting forms. That green growth on the exposed part of the wall is composed of three distinct species, belonging to as many different genera. The light green shaggy tufts are *Cladophora lætevirens;* the longer and darker green tubular fronds which grow amongst and overhang it, are *Enteromorpha compressa;* and the pendent flat fronds lower down are *Ulva Linza.* The delicate filmy purple plant which is so fragile that it is difficult to gather a perfect specimen, is *Porphyra vulgaris;* and closely applied to the wall like a crust, which any but a close observer would overlook, are dark-olive patches of *Ralfsia verrucosa,* and the harder, calcareous, purple fronds of *Melobesia.* Near the bottom of the wall are tufts of *Chondrus crispus,* at present in a young state, with their fronds imperfectly developed. In the pools, and on the larger Sea-weeds, are olive spots of what at first sight appears to be slimy mud, but which, on closer examination, will prove to be *Ectocarpus littoralis* and *E. tomentosus.* On the wall and boulders under the *Fuci,* which protect them from the sun and keep them moist when the tide is down, are lovely *Ceramiums* and delicate *Callithamnions;* each species with a colour and mode of growth peculiar to itself, by which, after it has once been thoroughly studied, it may be readily distinguished. Nor are these all. That smooth darkgreen coat on the wall, which has been hitherto overlooked, is a species of *Conferva;* and the shining purple spots are *Bangia fusco-purpurea,* which, though they appear insignificant to the naked eye, have an internal structure that will well reward the trouble of a microscopic examination.

This rough sketch from nature will, I trust, serve to show

how much may be found even in such a comparatively unproductive spot as that I have attempted to describe. I have purposely laid the scene in a locality far below the average of English watering-places as a collecting-ground for Sea-weeds, and the colours of the picture could be gradually heightened until they culminated in brilliancy on the glowing canvas that should worthily portray the varied products of the rich rocky shores of Devon and Cornwall, of Ireland, or of the Channel Islands.

I do not know that I have anything further to say on the subject of collecting, and I therefore pass to the process of preserving the specimens that may have been obtained.

This should be commenced as soon as possible after the collector returns from the shore, for many of the smaller species of the Red series begin to decompose directly they are taken from the water, and none but a few of the larger kinds, chiefly of the Olive series, will keep more than a day or two in a moist state.

The end to be attained is a dried specimen spread on and adhering to paper in such a manner as shall best display its natural appearance and characters. There are, however, some species that will not under any treatment stick to paper, and must, therefore, be simply washed in fresh water and dried between two pieces of rag.

With all the others, save a few that require exceptional manipulation which will be referred to hereafter under their respective descriptions, the *modus operandi* is somewhat in this wise :—The specimen should be first rinsed and thoroughly cleaned in a basin of water, specially provided for that purpose. It should then, if too large or too thick to be laid out entire, be reduced by the removal of all awkward or imperfect branches. This done, it should be transferred into a deep dish full of water, wherein a piece of paper of the requisite size has been previously placed. The operator should then support the paper in the water with his left hand, while with his right he carefully spreads the floating specimen over it, so as to exhibit the plant to the best advantage. Some species have a natural aptitude to lie flat, and with these the action of the water, aided by a little judicious manœuvring, will effect all that is required. Others are not so kindly, and call for the exercise of more or less skill, patience, and perseverance. Their branches and branchlets must be picked out and placed in position one by one with a fine knitting-needle, or some similar instrument; or, if they be very delicate, with the feather part of a quill pen or a camel's-hair brush. When the plant is properly spread on the paper, both should be very carefully removed from the dish and placed on some inclined surface for a few minutes, to allow the water to drain from them; but they must not be permitted to become too dry before they are put to press, or the paper will be apt to shrink unevenly.

The pressure must be proportioned to the size and texture of the specimens, and should be applied rather gradually. The best portable press that I am acquainted with consists of two pairs of beech-wood or mahogany bars, fitted with thumb-screws at both ends, and a few planed deal boards, rather less than half an inch thick, and about eighteen inches long by a foot broad. The bars should be made of well-seasoned wood, so that they may not readily warp, and should be sufficiently strong to bear the very considerable strain to which they will be subjected. The screws should be well finished, such as are used by cabinet-makers, and should be at least six inches long, so that several pairs of boards with specimens and rags may be pressed at once.

Various other modes of applying pressure will readily suggest themselves, such as books, weights, a tablecloth press, or such other means as may be within reach. The thin boards and a good supply of cotton or linen rags, no matter how well worn, to fit them, are all the apparatus absolutely indispensable. Whenever practicable, only one layer of specimens should be pressed between each pair of boards. Thus a pile for pressing should be formed in this sequence:—a board, a rag, a series of specimens sufficient to cover the rag without overlapping each other or the edges of the board, another rag, another board, and so on. If the specimens be thick and retain much moisture, a sheet or two of white blotting-paper may be added between the board and the rag, but no coarse or coloured paper that might indent or stain the laying-out paper, or destroy the texture of the specimens, must be used for this purpose.

In this mode, a day or two will suffice to dry the specimens enough to permit them to be transferred to some old heavy book to be finished off. As a rule, it is desirable to change the rags once or twice during the process of pressing; but this cannot always be done without disturbing the arrangement of the plant, and when there is any symptom that this will be the case, it is better to let the rags remain until the specimens be dry. If all the specimens between each pair of boards be of nearly uniform thickness, and be placed so that their branches lie in the same direction, they will dry more evenly, and there will be less risk of disturbing them when removing the rags, which should always be stripped off gently, commencing from the root.

I must not omit to refer to the paper to be used, first to lay out and subsequently to mount a collection. The best kind for the former purpose that I have seen is of French manufacture, wove, very white, thick, and not highly-sized. I obtained this in Jersey, but have not always been able to meet with it even there. The most suitable paper of English make is the very heaviest printing demy that can be procured. Dr. Cocks, in his 'Seaweed Collector's Guide,' states that a ream of the ordinary size, 17½ by 11 inches when folded once, should weigh about 34 pounds. He adds, "Having chosen a paper of this kind, I would strongly recommend that no other should be used." For mounting, cartridge-paper is best, and the size and quality must depend on the taste and the requirements of the collector. What is called elephant size, cut in half for the covers for each genus, and in quarters for the species, is a convenient and handsome form.

My own practice is to have two complete sets of papers,—one rough, with brown-paper genera, and whity-brown or newspaper specific cases; the other—my collection proper—of cartridge-paper throughout. I place all my specimens, in the first instance, in the former, and thence at my leisure transfer to the latter such as I may wish to keep permanently, leaving the remainder in the rough cases as duplicates.

I write the name and number of each genus, and of all the species that it contains, whether I possess specimens or not, outside the genera cases of both sets, so that I am always able to keep my whole collection in systematic order, and to refer readily to any part of it.

I shall insert a numbered, systematic list of genera and species at the end of this work, to aid those collectors who may wish to adopt the same plan.

Before I close this chapter, I must devote a few lines to a glance at my subject from one other point of view —the *cui bono* aspect, if I may so call it, of Sea-weeds.

In early times, when true science scarcely existed, and every product and phenomenon of Nature was valued according to its evident effect on man's real or supposed happiness, or in proportion to the gratification that it afforded, many things were deemed worthless simply because their uses were not understood, or their indirect action was overlooked.

This was the case with Sea-weeds, and in the rare instances that they are mentioned by ancient authors, the epithet 'useless' or 'vile' is always added.

The spread of scientific knowledge has reversed this verdict, and oceanic vegetation, which is almost wholly composed of Algæ, is admitted to be of vast importance in the economy of Nature. To it is entrusted the function of converting the mineral matter held in solution in the water into organic substances fit for food for the myriads of various animals that inhabit the sea. This it does, either directly or indirectly; the former in the case of those mollusks, fish, etc., that are vegetarians, the latter with those that are carnivorous,—so that we are in some degree indebted to the "vile alga" for every dainty dish of fish we eat. Had Horace, epicure that he was, thought of this, he would have written very differently on the subject, and would probably have immortalized the merits of the weeds which lack of knowledge of their virtues led him to despise.

Subsidiary to these general services there are many special purposes of agriculture, art, manufacture, and medicine to which certain kinds of Sea-weeds are applicable. On almost every coast where it is abundant, wrack, or vraich, a term applied in a general sense to Sea-weeds, is collected for manure or fuel, and the fertility of certain localities, as for instance the Channel Islands, is due to the facility with which this can be obtained. In Jersey there are very stringent laws as to the season when the vraich may be cut, and during the prescribed time every available man, woman, and conveyance is devoted to the service. The whole island reeks of vraich, and carts of all sorts and sizes, heavily laden with it, are to be met at every turn. The little island horses seem to thoroughly understand the business, and pick their way over the rocks with carts behind them in a manner perfectly astounding to the uninitiated beholder.

All the large weeds of the Olive series were formerly very extensively and profitably used in the manufacture of kelp, then one of the chief ingredients of glass and soap, but which has since been superseded by cheaper alkalies. The chemical substances iodine and mannite are obtained from Sea-weeds; and a species of *Gracilaria*, mixed in most cases with *Laurencia obtusa*, is used medicinally under the name of Corsican moss.

Chondrus crispus and some other species yield a gelatine said to be a remedy for consumption, and which is at any rate sufficiently good to be used in the preparation of blancmange. In China a cement of equal strength with gum and glue is derived from another species of *Gracilaria*. Several British species are reputed to be eatable, but none of them, with the exception, perhaps, of the *Porphyræ*, which yield 'sloke' or 'laver,' are very palatable. Chinese birds'-nest, which is composed of an Alga, is, however, a dainty delicacy when eaten in soup.

Knife- and whip-handles, and walking-sticks, are made from the stems of *Laminaria digitata*; and the *fronds* of *L. saccharina* frequently do duty as barometers. I need scarcely refer to sea-weed baskets and pictures, or to the more scientific books and maps of named specimens, which are often sold at bazaars, and by means of which considerable sums have been raised for rebuilding churches, and other similar purposes.

These are some of the uses of Sea-weeds, and I could enumerate others, did my space permit.

There is, besides, their beauty of colour, of structure, and of form, and to my mind this is by no means their least valuable attribute.

" For not to use alone did Providence
   Abound, but large example gave to man
   Of grace, and ornament, and splendour rich,
   Suited abundantly to every taste,
   In bird, beast, fish, winged and creeping thing,
   In herb and flower."

OLIVE-COLOURED SEA-WEEDS.—MELANOSPERMEÆ.

Order I. FUCACEÆ.

Fronds without joints, mostly large and of tough leathery texture. Spores in globular cavities, in the substance of the frond.

Genus I. SARGASSUM.

Frond branched, with a distinct stem bearing leaves. Air-vessels simple, on short stalks. Spore-receptacles small, generally in axillary clusters.

Upwards of one hundred species of this genus have been described. They are distributed over the warmer latitudes of both hemispheres, and are especially numerous in the Red Sea. The enormous masses of floating weed, which exist in the tropical part of the Gulf Stream, are composed of two or three species, chiefly of *S. bacciferum.* The early navigators called the Gulfweed *Sargazo,* or *Sea-lentils,* from the resemblance of its air-vessels, which they doubtless mistook for seeds, to the pods of the lentil. Hence the name of the genus. Two species are usually considered to belong to the English marine flora; but they are certainly not natives, and their only claim to be admitted is the fact that they have been occasionally picked up on our coasts.

Sargassum vulgare. Common Sea-lentils.

Stem smooth, slender, alternately branched. Leaves oblong-lanceolate, of variable breadth, serrated and ribbed. Air-vessels round, on short, flattened stalks. Spore-receptacles in the axils of the leaves.

This plant is found on the shores of the Southern States of North America, and on the coasts of Spain and Portugal, to which latter localities we are probably indebted for the few recorded so-called English specimens.

Sargassum bacciferum. Berry-bearing Sea-lentils, or Gulfweed.

Stem cylindrical, much branched. Leaves linear-lanceolate, serrated and ribbed, two to three inches long. Air-vessels on cylindrical stems, round, terminated by a short point. Fructification unknown.

The *Gulfweed* was probably known to the Phœnician navigators, certainly to Columbus, for he has left an account of it, and it is a curious fact that the position of the principal banks has not varied since his time. Floating in mid-ocean, this plant appears to have neither root nor spore, but to be propagated by means of branchlets broken from the mass by the action of the sea. Notwithstanding this apparently abnormal mode of growth, it is wonderfully prolific, and has been computed by Humboldt to cover an area of more than a quarter of a million of square miles of sea, or a space five times as large as that occupied by England.

Genus II. HALIDRYS.

Root a conical disc. Frond shrub-like, with branchlets having the appearance of leaves. Air-vessels pod-shaped, jointed, divided internally into cells, pointed at the apex. Fructification in terminal, tuberculated receptacles.

The name of this genus is derived from the Greek words *als*, the sea, and *drus*, an oak. Of the two species that are known, one has only been found in North American waters, and the other is common on all parts of the British coast and the adjacent shores of Europe.

Halidrys siliquosa. Podded Sea-oak.

Frond repeatedly pinnate. Air-vessels oblong, with a bristle-like point.

This Sea-weed is a perennial, and may be found at all seasons, in pools between high- and low-water mark. It varies in size from a few inches long in shallow water to three or four feet at greater depths.

Genus III. CYSTOSEIRA.

Root a conical disc. Frond shrub-like, having a woody stem with alternate branches. Air-vessels of one cell in the substance of the branches. Fructification at the ends of the branchlets.

The name *Cystoseira*, from the Greek words *kustis*, a bladder, and *seira*, a string, has reference to the arrangement of the air-vessels. The larger proportion of the twenty species which compose this genus are found in the Mediterranean Sea, and the remainder, with the exception of one or two which grow in America, frequent the shores of Europe. They appear to form the connecting link between the warmth-loving *Sargassa* and the *Fuci*, which delight in colder climates; and as in geographical position, so also in structure they are intermediate between these two genera, partaking of the characters of both.

Cystoseira ericoides. Heath-like Cystoseira.

Stem woody, short, with slender branches covered with awl-shaped spines or leaves. Air-vessels solitary, small, placed near the tops of the branches. Spore-receptacles cylindrical, with awl-shaped points.

The rock pools of the south and south-west coast of England and Ireland abound with this plant, which is also very plentiful in the Channel Islands. It is very rare in the northern districts, and has scarcely ever been found in Scotland. When seen growing under water it is beautifully iridescent, and may be easily distinguished by this character, which no allied species possesses.

Cystoseira granulata. Granular Cystoseira.

Stem covered with elliptical knobs, from which spring slender, repeatedly divided branches. Air-vessels two or three together. Spore-receptacles elongated.

The distinctive character of this plant is the knotted stems from which it takes its name. It is to be found in similar situations to *C. ericoides*, and is common in many localities.

Cystoseira barbata. Bearded Cystoseira.

Stem cylindrical, with small knobs bearing very slender, much divided branches. Air-vessels lance-shaped, not always present. Spore-receptacles small, with a bristle-like point.

The claim of this plant to a place among our native Sea-weeds is infinitesimally small, and the young collector need not, therefore, perplex himself by endeavouring to find it among his slender specimens of *C. granulata*, to which species he may safely refer all *Cystoseiræ* with knotted stems.

Cystoseira fœniculacea. Fennel-leaved Cystoseira.

Stem flattened; branches long, clothed with blunt spines. Air-vessels small, lance-shaped, placed below the forkings of the branchlets. Spore-receptacles very small, smooth, and without points.

The South of England and the Channel Islands are the best localities for this species, which is not found so far north as some of its kindred.

Cystoseira fibrosa. Fibrous Cystoseira.

Frond two to three feet long, very much branched; stem flattened, as thick as a swan's quill; branches slender. Air-vessels oval, large, imbedded in the lower part of the branchlets. Spore-receptacles very long, clothed with bristle-like fibres.

The only species with which this can be confounded is *C. ericoides*, from which it differs in its generally large size, more prominent air-vessels, and in the absence of the beautiful iridescence which appears on that plant when growing under water.

Genus IV. PYCNOPHYCUS.

Root fibrous, spreading. Frond a straight smooth shoot, entirely without branches for the first few inches, then forking, from rounded axils, into branchlets of unequal length which are again similarly divided. Air-vessels one-celled, in the substance of the frond, always obscure and frequently altogether absent. Fructification in cellular receptacles at the tips of the branchlets.

The Greek words *pycnos*, thick, and *phycos*, a sea-weed, furnish the not very descriptive name of this small but well-marked and widely distributed genus, whose geographical range extends from the south of Africa, along the Atlantic and Mediterranean shores of that continent, and Europe as far north as the west coast of Ireland.

Pycnophycus tuberculatus. Tubercular Pycnophycus.

Even when seen for the first time, this plant can scarcely be confounded with any other. Its smooth cylindrical fronds, bright olive-colour, and fibrous root, are unmistakeable characters. In consequence of its succulent texture it can only grow under water, and is, therefore, to be found in pools which are never left empty by the tide. For the same reason it shrinks much in drying; indeed the change is so great that the collector can scarcely believe the shrivelled, black, dried specimen to be identical with the handsome bright-coloured plant which he gathered.

Genus V. **FUCUS.**

Root a conical disc. Frond flat, with a midrib, or compressed without, forked. Air-vessels, one-celled, in the substance of the frond, not always present. Fructification in receptacles at or near the ends of the branchlets, or on independent stems or shoots.

Phycos was the Greek word for sea-weed, and early writers on the subject adopted *Fucus* as a generic name for nearly half the marine Algæ with which they were acquainted. Modern research has led to the construction of many new genera to which the bulk of the plants formerly belonging to this genus have been referred.

At present it contains only about a dozen species, which are all found in the northern latitudes of the Atlantic Ocean, either on the coasts of Europe or America, most of them being common to both continents. They grow in large numbers on rocks between the tide-marks, where their tough leathery texture and large size fit them to bear exposure to air and sun, and to afford shelter to more delicate species.

Fucus vesiculosus. Twin-bladder Wrack.

Frond flat, with a distinct midrib, not cut at the edges, varying in length from a few inches to two feet, or more. Air-vessels, when present, globular, generally in pairs, one on each side of the midrib. Spore-receptacles terminal, variable in shape, sometimes forked, full of mucus.

This is, perhaps, the most common British Sea-weed. It is abundant on all parts of our coasts, and is sometimes used as fodder for cattle, and more frequently as manure. It was formerly employed very largely, in combination with other species, in the manufacture of kelp, for the use of the glass-blowers and soap-boilers. In those days, a piece of rocky coast suitable for the growth, or, I might almost say cultivation, of Sea-weed was a valuable property; but improvements in the manufacture of alkalies have destroyed the trade of the kelp-maker, and with it the value of the rocky shore where it was conducted.

Fucus ceranoides. Horn Wrack.

Frond flat, with a midrib, not cut at the edges, about a foot long; main branches twice as wide as the branchlets springing from them, which are more distinctly forked. Air-vessels altogether absent. Spore-receptacles at the tips of the branchlets, small, spindle-shaped, sometimes forked.

This is not a common plant. It resembles *F. vesiculosus*, but may be readily distinguished by its narrow side-branches, its thinner texture, and the smaller quantity of saline matter which it contains. This latter peculiarity causes it to require less soaking, and to dry more quickly than other species of the genus.

Fucus serratus. Serrated Wrack.

Frond flat, forked, toothed at the edge, having a strong midrib, from two to four feet long, or more. Air-vessels, none. Spore-receptacles flat, at the ends of the branches.

This species is very common, and is easily identified. The width of the frond and the depth of its serratures vary considerably; but the general character of the plant is never lost.

Fucus nodosus. Knotted Wrack, or Seawhistles.

Frond flattened, without a distinct midrib, one to five feet long, a quarter of an inch to an inch wide; branches springing from slight projections, small at the base, and more or less pointed at the apex. Air-vessels very large, oval, formed in the main stem and branches. Spore-receptacles egg-shaped, on erect slender stalks, springing from projections on the branches, bright yellow when ripe.

This is the largest, toughest, and most rigid British *Fucus*. It grows nearer to low-water mark than any other species of the genus, and is usually more or less covered with *Polysiphonia fastigiata*. It may be easily recognized by its thick narrow frond and large air-vessels.

Fucus Mackaii. Mackay's Wrack.

Frond cylindrical, or slightly flattened, slender, much divided into forked branches. Air-vessels, when present, below the forkings of the frond, about half an inch long and one-fifth of an inch in diameter. Fructification in pendulous, lance-shaped receptacles on slender stalks, springing from the sides of the lower part of the branches.

The recorded British habitats of this rare plant are all in Ireland and Scotland. It was first found on the coast of Connemara, by Mr. Mackay, and is named after him. It has been considered to be only a variety of *F. nodosus;* but more careful examination has not confirmed this view, and it is now recognized as an established species. It grows in a round tuft, about a foot in diameter, somewhat like mistletoe, without a root, and not attached to anything, but resting on the sand or mud, or among rocks.

Fucus canaliculatus. Channelled Wrack.

Frond narrow, channelled on one side, rounded on the other, without either midrib or air-vessels, forked. Fructification in tubercular, forked receptacles, at the tips of the branches.

With the exception of the newly-discovered *F. anceps*, this is the smallest of our native *Fuci*, being rarely more than eight or nine inches high, and often only two or three. It grows very near high-water mark, and occasionally in situations where the spray alone reaches it, but in these positions it is very stunted.

Fucus anceps. Two-edged Wrack.

Frond repeatedly forked, flat in the lower part, tapering above, with an indistinct midrib. Air-vessels, none. Fructification in elongated, pointed receptacles, at the tips of the branches.

This plant is one of the most recent additions to our list of native Sea-weeds. It was discovered by Professor Harvey and Mr. N. P. Ward, at Kilkee, Ireland, in July, 1863, and was at first considered to be identical with the Linnæan species *F. distichus*. Subsequent examination and comparison with European specimens of the latter plant, established the fact that it was a distinct species, which had not been previously described, and it has, therefore, been duly installed as *F. anceps*, "Harvey and Ward." I dare hardly hope that many young collectors will be so fortunate as to secure this rarity; but those who may do so, will readily distinguish it from *F. canaliculatus*, which is the only British *Fucus* which it externally resembles, by its flat unchannelled stem, and long, tapering receptacles. The 'Journal of Botany' for 1863 contains a full description of this plant, by Mr. W. Carruthers, and a capital plate by Mr. Fitch.

Genus VI. HIMANTHALIA.

Root a small disc. Frond, in its young state, a pear-shaped sac, quickly becoming a button, at first flat-topped and hollow, then concave and solid. Fructification in long, linear, repeatedly forked receptacles, springing from the centre of the frond.

It is not easy to determine the exact derivation of the name of this plant. The Greek word *imas*, a strap, furnishes the first part, but it is doubtful whether the remainder is derived from *thalos*, a *branch*, or *als* or *thalassa*, both of which mean sea. The English name of Sea-thongs would be best translated by the latter word, and that probably is correct. This genus contains but one species.

Himanthalia lorea. Leather-thong Himanthalia, or Sea-thongs.

This is common on the Atlantic shores of Europe, and has been found on the coast of North America. Its chief peculiarity consists in the very large proportion that the spore-receptacles bear to the whole plant. These are commonly three or four feet long, and are said to attain a length of even twenty feet, while the little button frond is never more than an inch or an inch and a half high, and about two inches in diameter. Great diversity of opinion has been expressed as to the duration of this plant,—some writers asserting it to be annual, some biennial, and some even perennial. My own recent observations of a very large number of growing specimens in early spring lead me to the conclusion that the middle course is, as usual, the correct one, and that it is biennial. It certainly is not annual; but I am not prepared to say that it may not be perennial.

Order II. SPOROCHNACEÆ.

Fronds without joints. Spores attached to external, jointed filaments, either free or compacted together.

Genus VII. DESMARESTIA.

Frond linear, either thread-like or more or less flat, branched, with a single-tubed, jointed thread running through it, when young, bearing marginal tufts of branching fibres. Fructification unknown.

This genus is named in honour of the celebrated French naturalist, A. G. Desmarest. It does not contain many species, but is nevertheless widely distributed, and ranges over all the temperate and cold regions to the utmost limits of marine vegetation. The young and mature states of most of the species are very different, and puzzle those who see them for the first time. All the species not only decay very quickly after being gathered, but also cause any other weeds with which they may come in contact to do the same. They should, therefore, always be kept separate, and laid down as quickly as possible.

Desmarestia ligulata. Ligulate Desmarestia.

Frond flat, with an indistinct midrib, repeatedly pinnate; branches and branchlets opposite, tapering towards both ends.

Ccommon all round our coast. It varies chiefly in the width of the branches, which in some specimens is nearly one-third of an inch, while in others it is scarcely greater than in the filiform fronds of *D. viridis*, for which it is in consequence sometimes mistaken.

Desmarestia pinnatinervia. Pinnately-nerved Desmarestia.

Frond flat, leaf-like, with waved edges, from four inches to a foot or more long, and from an inch to two or even three inches broad, of a pale olive colour, and membranous texture, traversed throughout by a distinct mid-vein, from which spring exactly opposite side-veins.

This interesting plant is a comparatively recent addition to the British marine flora, and its title to be considered a distinct species is still in doubt. The first recorded British specimens were found by Mr. Lawers in Molville Bay, county Donegal, in 1853, and more recently it has been found at the Lizard, Cornwall, by Dr. Hermann Becker. I have not seen any of the Irish specimens, but I have now before me two of the plants collected by Dr. Becker. The latter appear to be smaller than those from Donegal, being only of the minimum dimensions above stated. They are entirely without branches, and altogether differ so completely in appearance from the forms of *Desmarestia ligulata* with which I am acquainted, that I have hesitated to adopt the suggestion that the plant is only an extravagantly wide form of that species, and prefer to retain it under a distinct name, pending the examination of a larger series of specimens than has been hitherto available.*

Desmarestia aculeata. Prickly Desmarestia.

Frond cylindrical at the base, becoming flattish in the upper part, much divided; branches and branchlets alternate, tapering at the base, when young of a bright green colour, tender substance, and fringed with slender threads; when old, brown, coarse, and covered with spines.

Desmarestia viridis. Green Desmarestia.

Frond cylindrical, filiform, much divided; branches and branchlets hair-like, opposite.

* Since writing the above, I have received from Dr. Becker a liberal supply of specimens in a fresh state, and a careful examination of these has fully convinced me that this plant is specifically distinct from *D. ligulata*, and is identical with the Continental *D. pinnatinervia*. I take this opportunity to thank Dr. Becker, both for the specimens and for the information concerning them, and to record the fact that he was the first, and, so far as I am aware, is the only discoverer of this interesting Sea-weed in an English habitat.

This is a very delicate and beautiful species, and is widely distributed in the higher latitudes, both north and south. Unlike most other plants, it appears to increase in luxuriance as it penetrates into the regions of the greatest cold.

Genus VIII. ARTHROCLADIA.

Frond thread-like, cellular, knotted, covered with whorls of jointed, bipinnate filaments, and traversed by a tube which is divided transversely into air-cells. Fructification in bead-like, stalked receptacles on the filaments.—ARTHROCLADIA, from the Greek *arthron*, a joint, and *klados*, a branch.

This genus contains only one species,

Arthrocladia villosa. Shaggy Arthrocladia, which is an annual, and grows in deep water. It is sparingly distributed on the shores of Europe, and has been found in North America. It is most frequent on the south coast of England and in Jersey. I well remember the delight of finding fine specimens of this plant and of the allied species, *Sporochnus pedunculatus* and *Desmarestia viridis*, all together, and for the first time, in Grève d'Azette, at the back of Elizabeth Castle, St. Helier, and I trust that some at least of my readers may be equally fortunate.

Genus IX. SPOROCHNUS.

Frond thread-like, solid, composed of two kinds of cells, those at the axis and circumference very small, those intervening much larger. Fructification stalked, oblong receptacles, crested with tufts of slender, jointed fibres, and composed of branched spore-bearing filaments arranged round a slender axis.—SPOROCHNUS, from the Greek *sporos*, a seed, and *chnoos*, wool, in reference to the tufts of fibres which adorn the spore-receptacles.

This genus contains four or five species, all of which are natives of temperate climes. One only is found on our coasts, and that but sparingly.

Sporochnus pedunculatus. Pedunculated Sporochnus.

Frond from a few inches to a foot or more long, pinnately branched, of a pale olive-colour and tender texture; branches long, simple, set on at right angles to the stem. Spore-receptacles elliptical.

This is a very beautiful plant, and is easily distinguished by its slender, simple stem, and its thread-like, tuft-covered branches.

Genus X. CARPOMITRA.

Frond flat, forked, mid-ribbed. Fructification in mitre-shaped receptacles, at the ends of the branches, composed of whorls of branched, spore-bearing filaments arranged round a vertical axis.—CARPOMITRA, from the Greek *karpos*, a fruit, and *mitra*, a cap.

The number of species of this genus is very limited, and only one occurs in a northern latitude.

Carpomitra Cabreræ. Cabrera's Carpomitra.

Root a small tuber. Frond about six inches high, forkedly branched, mid-ribbed; branches erect, linear, narrow, flat, constricted at intervals.

Dr. Harvey, in the 'Phycologia Britannica,' writes of this plant:—" Specimens having never been found but once, and then only washed on shore, we may be allowed to entertain the fear that this interesting plant is not truly the growth of our shores, but wafted hither, as extra-European productions sometimes are, by the force of currents." Since this was written, specimens have been dredged at Plymouth, and others picked up in Jersey. I do not hesitate, therefore, to express my opinion that this rare and interesting species is entitled to a place among our native Sea-weeds.

Order III. LAMINARIACEÆ.

Fronds without joints. Spores superficial, either forming cloud-like patches, or covering the whole surface of the frond.

Genus XI. ALARIA.

Root fibrous. Frond stalked, with a strong midrib throughout its length; stem winged with ribless leaflets, which contain the fructification. Spores pear-shaped, arranged in oblong sori.—ALARIA, from the Latin *ala*, a wing.

There are about five known species of *Alaria*, which are distributed over the northern shores of the Atlantic and Pacific oceans.

Alaria esculenta. Eatable Alaria, or Badderlocks.

Frond elongated, lance-shaped, entire; rib narrow; leaflets linear-oblong or wedge-shaped.

The midrib of this plant is eaten in Ireland and Scotland, and in some of the adjacent islands; hence the specific name *esculenta*.

Genus XII. LAMINARIA.

Frond leathery or membranous, flat, with a stalk, but no

mid-rib. Fructification imbedded in the frond in spots.—
LAMINARIA, from the Latin *lamina*, a thin plate.

There are several species of this genus, which are widely distributed both in hot and cold climates. They grow generally at and beyond low-water mark, extending into depths of many fathoms. With one or two exceptions, all the species are of large size, and some of them attain a height of twenty feet. They grow very rapidly, and renew the leafy part of their fronds every year. The new leaf is formed between the top of the stalk and the base of the old leaf, which gradually decays and gives place to its successor. This process is very well exhibited in small specimens of *L. digitata*, and the young collector should obtain a few plants, showing the two fronds in various stages of growth.

Laminaria digitata. Fingered Laminaria, or Tangle.

Stem solid, cylindrical, tapering upwards, of varying length; frond leathery, roundish-oblong, when quite young, entire, rapidly becoming deeply cleft into several strips.

Small specimens of this species may be gathered in the deeper tide-pools near low-water mark. The larger are cast on shore after storms. It is impossible to preserve the latter, but they should always be examined, as many rare small Algæ, which delight in deep water, grow on their stems, and can only be collected in this manner.

Laminaria digitata, var stenophylla.

This plant is figured by Professor Harvey in the 'Phycologia Britannica,' and he expresses a doubt as to whether it is not entitled to rank as a distinct species. The stem is longer and less rigid, and the whole frond is much more more limp and slender than the ordinary form of *L. digitata.* There is also a difference in the root.

Laminaria bulbosa. Bulbous Laminaria.

Stem flat, with a waved margin, and having a hollow bulb or tuber just above the root; frond oblong, cleft into several strips.

The distinctive character of this plant is the bulb from which it takes its name, and which is to be found in specimens of all ages.

Laminaria longicruris. Long-stalked Laminaria.

Stem eight to ten feet long, solid at the base, becoming thicker and hollow in the middle, and thence tapering upwards; frond from six to eight feet long, and from two to three feet wide, oval, waved at the edge, rounded at the top.

This is another of the waifs and strays of foreign climes which has been occasionally found on our shores in a very worn and imperfect state, and which has, I fear, but small claim to be admitted to a place among our native Sea-weeds. Dr. Harvey has included it in his 'Phycologia Britannica,' and I follow his example. It is abundant on the coast of North America.

Laminaria saccharina. Sugared Laminaria.

Stem cylindrical, slender; solid; frond leathery or almost membranous, lanceolate, entire.

This is a very well-known common species, and is probably familiar to many of my readers as the cottager's weather-teller. It is very apt to sport, and many very interesting, abnormal forms may be obtained by those who will take the trouble to look for them.

Laminaria Phyllitis.   Hart's-tongue Laminaria.

Stem short, slightly flattened; frond linear-lanceolate, membranous, entire.

Some doubts are still admitted to exist as to whether this is a distinct species, or only a very slender variety of *L. saccharina.* The resemblance between the two is close, but after careful examination Dr. Harvey is of opinion that *L. Phyllitis* should be retained as a species.

Laminaria fascia.   Ribbon Laminaria.

Stem very short and slender; frond membranous, of very varying width and shape, from linear to broadly oblong, entire.

This is not a common species, but may be found in many localities all round our coasts. It is the smallest of the British Laminariæ, seldom attaining a height of more than nine or ten inches.

Genus XIII.   CHORDA.

Root shield-shaped. Frond a simple cylindrical tube, divided internally by transverse membranes. Fructification a layer of obconical spores, elongated at the base, spread over the external surface of the frond.—CHORDA, from the Greek *chorde,* a string or cord.

The geographical distribution of this genus is very general, and the individuals are numerous on all the northern shores of the Atlantic and Pacific Oceans.

Chorda filum. Thread Chorda.

Frond cartilaginous, slimy, tapering from the middle to each end, varying in length from a few inches to thirty or even forty feet. When young, covered with slender fibres, which are worn off as the plant matures; occasionally the tips of the fronds become inflated, and may be seen floating above the surface of the water, but specimens in this state are not common.

Chorda lomentaria. Constricted Chorda.

Frond a simple membranous tube, slender at the base, constricted at distant intervals, so as to give it the appearance of being pointed, tapering at the tip.

The differences between this and the preceding species are so great that some writers are doubtful whether both should be included in the same genus. *C. lomentaria* occurs frequently on our coasts. Its fronds usually grow in tufts, are thin in substance, free from slime, and are not more than a foot long, frequently only a few inches. I gathered specimens near Whitby, Yorkshire, of a very small variety which was growing abundantly on the rocks near high-water mark. The fronds were only two or three inches long, not constricted, and, when growing, had the appearance of tufts of stunted grass.

Order 4. DICTYOTACEÆ.

Fronds without joints. Spores superficial, disposed in definite spots or lines.

Genus XIV. CUTLERIA.

Root covered with woolly fibres. Frond flat, of thickish substance, expanding upwards, irregularly cleft or forked. Fructification of two kinds on separate plants, dotted over the whole surface of the frond:—1, tufts of stalked cells, containing spores; 2, minute branches of fibres, bearing linear, dotted, sessile antheridia.—CUTLERIA, named after Miss Cutler, a lady whom every student in marine botany should delight to honour.

The number of known species is only four or five. They are distributed over the southern shores of Europe.

Cutleria multifida. Many-cut Cutleria.

Frond varying in length from two or three inches to a foot or more, wedge-shaped, divided into irregular, forked branches, which spring from very acute axils, and have tufts of delicate, jointed fibres on their pointed tips. Fructification in dots, attached to a network of branching, jointed fibres spreading over the whole frond.

This very beautiful and interesting plant has been found in many British localities, but does not appear to be abundant in any of them. It is annual, and grows on rocks which are not left bare, even by the lowest tides. The collector must, therefore, either dredge for it or be content with such specimens as may be cast on shore, and these are not numerous, except after storms. Externally *Cutleria multifida* differs from *Dictyota dichotoma* in having a more wedge-shaped and less divided frond, which, when fertile, is covered all over with dots of fructification.

Genus XV. HALISERIS.

Frond membranous, flat, forked, mid-ribbed. Fructification, clusters of egg-shaped spores, generally arranged in rows running lengthwise of the frond.—HALISERIS, from *als*, the sea, and *seris*, endive.

There are several species, all growing in warm climates, and only one extending so far north as our southern and western shores.

Haliseris polypodioides. polypody-like Sea-Endive.

Fronds forked, entire at the edge, growing in tufts, from six inches to a foot high, and having a peculiar pungent smell. Fructification in a line of dots on each side of the midrib.

This is a rare and beautiful species, and one that cannot be mistaken for any other. It grows in deep water, or only just above extreme low-tide mark. It must, therefore, be sought during the lowest spring tides which occur at new or full moon. As it is prone to decay rapidly, and to cause other weeds with which it may be in contact to do the same, it should be kept separate, and laid down as quickly as possible.

Genus XVI. PADINA.

Root coated with woolly fibres. Frond ribless, leathery, flat, fan-shaped, marked with concentric lines, fringed with jointed fibres, and rolled inwards at the top. Fructification in lines running along the markings of the frond beneath the epidermis.—The derivation of PADINA has not been traced.

Padina pavonia. Peacock's-tail Padina.

The one species to which this genus is now restricted is widely distributed in warm latitudes, and is especially abundant in the Mediterranean. It grows in the deeper tide-pools, some distance above low-water mark, and when seen in a luxuriant state under water is beautifully iridescent, and has very much the appearance of miniature peacocks' tails. It is found in several localities on the south coast, and in Jersey. Specimens usually require to be freed from sand and other foreign matters before being laid down, and the greatest care must be taken not to remove the fibrous fringe or to destroy the epidermis during this operation.

Genus XVII. ZONARIA.

Root coated with woolly fibres. Frond flat, ribless, leathery, membranaceous, fan-shaped, entire or cleft vertically into radiating segments, marked with indistinct concentric lines. Fructification roundish or linear clusters of spores formed beneath the cuticle of the frond, and bursting through at either surface.—ZONARIA, from the Greek *zone*, a zone or girdle.

This genus contains several species, which, with the exception of the two following, are confined to warm regions.

Zonaria collaris. Collar Zonaria.

Frond composed of two distinct parts; the lower or primary procumbent, leathery, orbicular, sometimes divided into lobes, with a dense woolly coating on its under surface, by which it adheres to the rocks; the upper or secondary, springing from the lower, cup-shaped, membranous, the edge notched at distant intervals, and fringed with delicate, jointed fibres. Fructification not hitherto discovered on Jersey specimens.

The claim of this beautiful species to be included in our flora rests at present only on Channel Island specimens. Professor Harvey first received it from Miss Turner, who picked it up in 'Granville' (Grouville?) Bay; and I have since seen several specimens, of the secondary part of the frond only, from the same or other Jersey localities.

Zonaria parvula. Little Zonaria.

Frond procumbent, membranaceous, variously lobed, spreading over the rock in circular patches, attached by numerous fibres issuing from its lower surface. Fructification not observed on British specimens.

Professor Harvey writes:—"This is not an uncommon plant on various parts of our coasts, though frequently overlooked, owing to its hiding in crevices, or creeping through the much-branched, stony Nullipores."

Genus XVIII. TAONIA.

Root covered with woolly fibres. Frond flat, ribless, networked, irregularly cleft, the surface cellules equally distant, those in the ultimate divisions of the frond in parallel or only slightly divergent series. Fructification in wavy concentric lines on both surfaces of the frond.—TAONIA, from the Greek *taon*, a peacock.

In the 'Phycologia Britannica,' *T. atomaria* is included in the genus *Dictyota*, but in his subsequent works Professor Harvey has adopted the name Taonia. In his 'Nereis Boreali-Americana,' he speaks of this genus as formed for the reception of the old *Ulva atomaria*, which has been variously referred to *Zonaria*, *Dictyota*, and *Padina*; and we will therefore hope that at last this much-vexed plant has found a local habitation and a permanent name.

Taonia atomaria. Banded Taonia.

Fronds from four inches to a foot high, and from half an inch to three inches wide, growing in clusters. Fructification arranged in wavy bands across the frond.

This fine species is found in several localities on the

British coasts and in the Channel Islands, but seldom in great abundance. It is annual, and is in perfection about July.

Genus XIX. DICTYOTA.

Root coated with woolly fibres. Frond flat, ribless, membranous, forked, networked; the surface cellules minute, equidistant, converging at the tips of the frond, which end in a single cell. Fructification, roundish clusters of spores scattered over the whole frond beneath the cuticle, through which they burst at maturity.—DICTYOTA, from the Greek *diktuon*, a net, in allusion to the network on the surface of the fronds.

This genus contains several species, which are distributed over very various latitudes, from the tropics to the North Sea.

Dictyota dichotoma. Forked Dictyota.

Frond regularly forked, from four inches to a foot long, and from an eighth of an inch to half an inch wide, wedge-shaped at the base then linear. Fructification in dots scattered over the frond, but not extending quite to either margin.

This species is very common, and varies much according to the conditions under which it grows. In still, deep water, the fronds become broad and robust, and *vice versâ*, in shallow water and exposed positions, they are small and narrow.

DICTYOTA, DICHOTOMA var. INTRICATA.—This form, which has been described by some authors as a distinct species, has fronds of a dark-brown colour, thick substance, very narrow, elongated, spirally twisted and tangled.

Genus XX. STILOPHORA.

Root a small disc. Frond cylindrical, branched, traversed by a cavity, which increases in size as the plant matures. Fructification, convex, wart-like clusters of spores studded over the surface of the frond.—STILOPHORA, from the Greek *stile*, a point, and *phoreo*, to bear.

The species of this genus are distributed on the shores of the Atlantic Ocean, and of the Mediterranean and Baltic Seas.

Stilophora rhizodes. Root-like Stilophora.

Frond almost solid, much divided into narrow branches bearing scattered, forke: branchlets. Fructification thickly studded over the whole plant; fibres of the sori simple; spores attached to them.

Stilophora Lyngbyæi. Lyngbye's Stilophora.

Frond tubular, with spreading branches forking from wide rounded axils, and bearing scattered hair-like branchlets with very slender forked tips. Fructification arranged round the frond in transverse lines at short intervals; fibres of the sori branched or forked, spores attached to them.

Professor Harvey has figured both these species in the 'Phycologia Britannica,' but expresses a doubt whether the differences between them do not depend on the relative depths of water in which they grow, *S. rhizodes* being found within the tide-marks, while *S. Lyngbyæi* always grows in deep water. At the same time, he describes and figures a considerable variation in the fructification of the two plants to the effect that: in *S. rhizodes* the fibres are simple, and the spores fixed to the surface of the frond; while in *S. Lyngbyæi* the fibres are branched or forked, and the spores are attached to them. Subscquently, however, he appears to have altered this opinion, and in his 'Nereis Boreali-Americana' he describes and figures the spores of *S. rhizodes* as being attached to the bases of the fibres.

Genus XXI. DICTYOSIPHON.

Root a small disc. Frond thread-like, tubular, branched. Fructification, naked spores scattered over the frond either singly or in clusters.—DICTYOSIPHON, from the Greek *dictuon*, a net, and *siphon*, a tube.

Dictyosiphon fœniculaceus. Fennel Dictyosiphon.

Frond very much divided into long, slender branches, bearing awl-shaped branchlets.

When young the fronds of this species are of a greenish-olive colour, and are covered throughout with jointed fibres; mature specimens are brown. This plant is annual, and grows abundantly in tide pools during spring and summer.

Genus XXII. STRIARIA.

Root a small disc. Frond tubular, of thin substance, branched. Fructification small clusters of spores arranged in zones round all parts of the frond.—From the Greek *strix*, or Latin *stria*, a fluting, in allusion to the arrangement of the spores in transverse lines.

Striaria attenuata. Tapering Striaria.

Fronds much branched, growing in tufts, three inches to a foot high; branches and branchlets nearly opposite, tapering at each end.

This species delights in quiet, sheltered situations, and grows at a considerable depth. It is annual, and in perfection in summer.

Genus XXIII. PUNCTARIA.

Root shield-shaped. Frond membranous, simple, flat, without a mid-rib. Fructification in minute dots scattered over the surface of the frond.—PUNCTARIA, from the Latin *punctum*, a dot.

This genus contains several species which are distributed on the shores of the Atlantic and in the Mediterranean Sea. Three of them are moderately common on our coasts.

Punctaria latifolia. Broad-leaved Punctaria.

Frond six inches to a foot long, and one to two inches broad, oblong, tapering abruptly at the base, irregularly waved and notched at the edge, of a thin membranous substance.

This species grows on rocks and Algæ in tide-pools. It is annual, and in perfection in summer. I recently gathered fine specimens in April at the back of Elizabeth Castle, Jersey.

Punctaria plantaginea. Plantain Punctaria.

Frond four inches to a foot long, and from a quarter of an inch to two inches broad, lance-shaped, gradually tapering at the base, of a leathery, membranous substance, and dark olive-brown colour.

Dr. Harvey, in his 'Nereis Boreali-Americana,' suggests that this is probably the type species, and that *P. latifolia* and *P. tenuissima* are only varieties. Possibly future botanists may so decide, but at present I prefer to retain the three species as figured and described in the ' Phycologia.'

Punctaria tenuissima. Slender Punctaria.

Frond two to eight inches long, and from one-tenth to

one-third of an inch wide, of exceedingly thin texture and pale olive-brown colour, growing in dense tufts round the leaves of *Zostera marina*, etc. No fructification has been hitherto observed.

Genus XXIV. ASPEROCOCCOUS.

Root shield-shaped. Frond a membranous tubular sac, sometimes compressed, but always consisting of two separable membranes. Fructification in minute clusters of spores scattered over the frond.—Asperococcus, from the Latin *asper*, rough, and the Greek *coccos*, a seed.

Various species of this genus occur on the Atlantic coasts of Europe, Africa, and America, and in the Mediterranean. Three are indigenous to the British Isles.

Asperococcus compressus. The compressed Asperococcus.

Frond flat, from four to eighteen inches long, and from a quarter of an inch to an inch and a half broad, linear-lance-shaped, with an obtuse apex and short stem. Spore-clusters oblong, thickly scattered over the frond.

This plant is the connecting link between the genera *Punctaria* and *Asperococcus;* but while possessing some of the characters of the former, its tubular, although much compressed fronds connect it more closely with the latter.

Asperococcus Turneri. Turner's Asperococcus.

Fronds growing in tufts, from a few inches to two or three feet long, and from half an inch to four inches in diameter, tubular, inflated, obtuse at the points, contracted into short stems at the base, of thin substance, and very variable shape. Spore-clusters minute, roundish, thickly scattered over the frond.

This species delights in still, muddy bays, where it sometimes attains the extreme size mentioned above. It grows on rocks, and on *Zostera* and algæ.

Asperococcus echinatus. Prickly Asperococcus.

Fronds growing in dense tufts, from two to eighteen inches long, and from one-tenth to half an inch in diameter, tubular, tapering gradually at the base, then continuing of the same diameter, and terminating in rounded or acute points. Spore-clusters small, rough, spread thickly over the whole frond. When young, the frond is covered with hair-like fibres.

Genus XXV. LITOSIPHON.

Frond cylindrical, cartilaginous, without branches, formed of concentric layers of cells. Fructification, naked spores scattered over the frond, either singly or in clusters.— LITOSIPHON, from the Greek *litos*, slender, and *siphon*, a tube.

The two species which constitute this genus are both parasitical, and are pretty generally distributed wherever the plants on which they grow occur.

Litosiphon pusillus. Small Litosiphon.

Fronds green, tufted, thread-like, from two to four inches long, with a networked surface, covered with minute, jointed fibres.

This plant grows on *Chorda filum* and *Laminaria bulbosa*, frequently clothing the frond of the former and the stem of the latter for a considerable distance. Although called the Small Litosiphon, it is many times larger than its companion species.

Litosiphon laminariæ. Laminaria Litosiphon.

Fronds brown, tufted, from a quarter to half an inch long, slightly tapering at the base, rounded at the apex, which is clothed with minute fibres.

This species is parasitic on *Alaria* (formerly *Laminaria*) *esculenta*, hence its name of *laminariæ*, which I should like to change to *alariæ* so as to denote correctly the plant on which it grows. There is, however, a strong and well-founded objection to any alteration of specific names, to which I gladly defer.

ORDER V. CHORDARIACEÆ.

Fronds gelatinous or cartilaginous, composed of interwoven, vertical and horizontal filaments. Spores attached to the filaments, and immersed in the frond.

Genus XXVI. **CHORDARIA.**

Frond cylindrical, branched, having a cartilaginous axis, surrounded by a periphery of club-shaped, whorled threads, and long, slender, gelatinous fibres. Fructification, egg-shaped spores arranged among the threads of the periphery.—CHORDARIA, from the Greek *chorde*, a string.

Representatives of this genus are widely dispersed in various latitudes, being found on the shores of Southern Africa, North America, and Europe, extending as far north as Iceland.

Chordaria flagelliformis. Whip Chordaria.

Fronds from a few inches to two feet or more long, of uniform thickness throughout; stem simple, bearing numerous, long, thread-like branches, almost destitute of branchlets.

This plant is annual. It grows on rocks, between high- and low-water mark, and is moderately abundant on our coasts.

Chordaria divaricata. Spreading Chordaria.

Frond irregularly divided; branches spreading, flexuous, their upper parts having short, forked branchlets.

The only British habitat hitherto recorded for this species is Belfast Lough.

Genus XXVII. MESOGLOIA.

Frond thread-like, much branched, gelatinous, having an axis composed of interlaced, longitudinal threads, covered with gelatine, and a periphery of radiating, forked filaments tipped with clusters of club-shaped, beaded fibres. Fructification, obovate spores developed among the fibres at the tips of the filaments of the periphery.—MESOGLOIA, from the Greek *mesos*, the middle, and *gloios*, glutinous.

The species which compose this genus are mostly inhabitants of cold climates. They are all extremely gelatinous, and may be distinguished by this character from all other Olive Sea-weeds, none of which possess it in anything like the same degree.

Mesogloia vermicularis. Worm-like Mesogloia.

Fronds tufted, from one to two feet high, gelatinous, limp, and elastic, having a main stem of unequal thickness, tapering at each extremity, and bearing numerous side-branches, which are more or less pinnate or forked.

The colour of this species is a brownish-olive. The spores have no stems.

Mesogloia Griffithsiana. Mrs. Griffiths's Mesogloia.

Fronds from a foot to a foot and a half high, with a slender stem of equal thickness throughout, bearing long, alternate, thread-like branches, either naked, or with a few short branchlets set on, at nearly a right angle to the branch, at distant intervals.

This species is of a pale olive-green colour, and its spores have short stems. It was discovered by Mrs. Griffiths, whose name it appropriately bears. The recorded habitats are chiefly on the south-west coast, and in Ireland, and even in these localities it is not of frequent occurrence.

Mesogloia virescens. Pale-green Mesogloia.

Frond from a few inches to a foot long, with a slender, cylindrical, undivided stem of the same thickness throughout, and bearing numerous side-branches, which are thickly studded with short simple branchlets.

The colour of this plant is a pale yellowish, or greenish-olive. Its spores are without stems, and the filaments of the periphery consist of a short stem, bearing a bundle of forked, beaded branches.

Genus XXVIII. LEATHESIA.

Frond tuberous, composed of jointed, forked threads. Fructification, oval or pear-shaped spores hidden among the outer series of filaments.—LEATHESIA, named in honour of the Rev. G. R. Leathes, a British naturalist, who discovered the first specimens which were described.

This genus is widely distributed in various latitudes.

The external appearance of the plants which compose it does not at all betoken the very beautiful structure which will be revealed when they are dissected and examined under the microscope.

Leathesia tuberiformis. Tuber-shaped Leathesia.

Fronds growing singly on other sea-weeds, or in clusters, on rocks; when young, filled with fibres, becoming hollow sacs as they mature, varying from a quarter of an inch to an inch and a half in diameter.

Leathesia crispa. Crisped Leathesia.

"Frond solid; axial threads very densely crowded, simple, dichotomous, with very long joints; peripheric ramuli club-shaped, incurved, or arcuate, submoniliform, the joints as long as broad. Spores pear-shaped.—On *Chondrus crispus*."

This species is not included in the 'Phycologia Britannica,' but was described by Dr. Harvey in the 'Natural History Review' for 1857. I have not seen a specimen, and have therefore taken the above description *verbatim* from Dr. J. E. Gray's 'Handbook of British Algæ.'

Leathesia Berkeleyi. Berkeley's Leathesia.

Frond flattened, solid, the fibres densely packed.

This plant has been found on the south-west coast of England, and in considerable abundance on the west coast of Ireland. It probably grows in many other localities, but its dark-brown colour, frequently of nearly the same shade as the rock to which it adheres, causes it to be overlooked by any but a very careful observer.

Genus XXIX. RALFSIA.

Frond leathery, crust-like, composed of jointed, closely packed, vertical threads, spreading in irregular patches over the surface of the rock to which its lower surface is attached. Fructification in warts scattered over the upper side of the frond.—RALFSIA, named in honour of John Ralfs, Esq., a well-known botanist,

Ralfsia verrucosa. Warty Ralfsia.

Fronds from one to six inches in diameter, of a dark brown colour, marked with concentric lines.

This very peculiar plant, which is called *Ralfsia deusta* in the 'Phycologia Britannica,' has more the appearance of a lichen than of a sea-weed. It is common on the shores of the North Atlantic Ocean.

Genus XXX. ELACHISTA.

Fronds parasitical, composed of two kinds of jointed threads; the inner, or axial series, forkedly branched, and combined into a tubercle; the outer, or peripheric, not branched, coloured, radiating from the tubercle. Fructification, pear-shaped spores attached to and hidden among the inner threads.—ELACHISTA, from the Greek *elachista*, the least.

All the species of this genus are insignificant, and very liable to be overlooked, and in consequence their geographical distribution has not been very accurately determined. Certain of them are known to be common on all the shores of Europe, and on the coast of North America, and they are probably equally abundant in other localities, indeed, wherever the plants on which they are parasitical are found. The beginner will not easily make out the specific characters of these minute plants, and it will be useless for him to attempt to do so without the aid of a microscope.

Elachista fucicola. The Fucus-inhabiting Elachista.

Tufts brush-like; tubercle hemispherical; the outer threads about an inch long, attenuated upwards, the intervals between the joints once and a half to twice as long as the threads are broad. Spores at first oblong, becoming attenuated as they mature, attached to the terminal fibres of the tubercle.

This is the largest, best known, and most abundant species of the genus. It grows on *Fucus serratus* and *F. vesiculosus*.

Elachista flaccida. Flaccid Elachista.

Tufts brush-like; tubercle small; outer threads half an inch long, tapering suddenly at the base, and gradually from the middle to the tip, the intervals between the joints in the lower half of the threads not quite half as long as the threads are broad, increasing towards the tip until they are as long or longer than the diameter of the threads.

This species is usually found growing on *Cystoseira fibrosa*.

Elachista stellulata. Starred Elachista

Tufts very minute, star-like; tubercle composed of large cells; outer threads very short, tapering at the base, the intervals between the joints twice as long as the threads are broad. Spores obovate, short-stalked, lodged among the outer fibres of the tubercle.

The only recorded specimens of this species are those originally discovered by Mrs. Griffiths, growing on *Dictyota dichotoma*. The tufts are so extremely minute as to be hardly discernible even with a lens, and it is, therefore, highly probable that they frequently occur on *Dictyota*, but are overlooked, or are thought to be the fructification of that plant.

Elachista scutulata. Little-shield Elachista.

Tubercle oblong, varying in length from half an inch to two or three inches, composed of a dense mass of forked fibres; outer threads short, of the same thickness throughout, rounded at the top, the intervals between the joints about three times as long as the threads are broad. Spores oblong, rounded at both ends, attached by long, jointed stalks.

This species grows on the thongs of *Himanthalia lorea*, and is abundant during summer and autumn. It is a very beautiful object for the microscope.

Elachista curta. Short Elachista.

Tufts minute; outer threads slender at the base, gradually increasing in size towards the tips, which are blunt; the intervals between the joints as long as the threads are broad. Spores on comparatively long stalks.

Dr. Harvey in his 'Phycologia' states that he has never found this plant, and expresses a doubt whether it be really distinct from *E. fucicola*. It is said to grow on the same *Fuci* as that species. I have never seen it myself.

Elachista pulvinata. Cushioned Elachista.

Tufts very minute, globose; outer threads tapering from the middle to each end, the intervals between the lower joints, three to four times as long as the threads are broad, gradually decreasing towards the tip, where they are very short. Spores linear-oblong, without stalks, at the base of the threads.

This species grows on *Cystoseira ericoides*, and has not hitherto been very frequently found. It is called *E. attenuata* in the 'Phycologia Britannica.'

Elachista velutina.   Velvety Elachista.

Patches velvet-like, spreading, of indefinite shape ; outer threads short, of the same thickness throughout, the intervals between the joints about once and a half as long as the threads are broad. Spores egg-shaped, dark olive-coloured, with wide transparent margins, borne on short slender stalks.

This species is found on the thongs of *Himanthalia lorea*, and sometimes in company with *E. scutulata*, which it very much resembles. It also grows on *Fuci*.

Elachista Grevillei.   Greville's Elachista.

Tufts brush-like ; threads cylindrical, slender, elongated, the intervals between the lower joints short, those between the upper rather longer.

This species is not mentioned in the 'Phycologia,' but was described by Dr. Harvey in the 'Natural History Review' for 1857. It grows on *Cladophoræ*.

Genus XXXI.   MYRIONEMA.

Fronds minute, parasitical, composed of two series of jointed threads, the inner branched and spreading over the surface to which the plant adheres ; the outer simple and springing at right angles from the expansion formed by the inner, both series united by gelatine, into a compact, cushion-like mass. Spores oblong, affixed either to the erect or to the spreading threads.—MYRIONEMA, from the Greek *murios*, numberless, and *nema*, a thread.

This is another genus of minute parasites, smaller even than the *Eluchistæ*.

Myrionema strangulans. Choked Myrionema.

Patches at first only dark-brown spots, consisting of the spreading threads, afterwards the erect threads are developed, and the patches become convex and gelatinous. Spores on short stalks springing from the spreading threads.

This species grows on *Ulvæ* and *Enteromorphæ*.

Myrionema Lechlancherii. Lechlancher's Myrionema.

Patches orbicular; the spreading threads, which are developed first, as in the last species, radiate from a centre, the erect threads are densely set together, and are longest in the middle of the patch, whence they gradually become shorter as they approach the side. Spores on long stalks.

This species grows on *Rhodymenia palmata*, and sometimes on *Ulva*. In autumn the fronds of the former are frequently covered with dots, which appear to the naked eye to be but the first symptoms of decay, but which when examined under a moderately powerful microscope exhibit the beautiful structure of this minute parasite.

Myrionema punctiforme. Dot-like Myrionema.

Patches very minute, globose; spreading threads forming a small base, from which the erect threads radiate. Spores linear-egg-shaped, attached to the bases of the erect threads.

But few specimens of this plant have been recorded, probably on account of its extreme minuteness. It grows on *Ceramium rubrum* and other Red Sea-weeds.

Myrionema clavatum. Club-shaped Myrionema.

The only recorded specimens of this species were discovered by Captain Carmichael, of Appin, who thus describes it :—" Very minute, rather convex; filaments clavate, mostly bifid; spores obovate, pedicellate, affixed to the filaments."

ORDER VI. ECTOCARPACEÆ.

Fronds jointed, thread-like. Spores attached to the branchlets or embedded in their substance.

Genus XXXII. CLADOSTEPHUS.

Frond cylindrical, branched, stem without joints, branches covered with whorls of short, jointed branchlets. Fructification, stalked spores on accessory branchlets, and in a mass called a *propagulum* at the tips of certain of the ordinary branchlets.—CLADOSTEPHUS, from the Greek *clados*, a shoot, and *stephos*, a crown.

Cladostephus verticillatus. Whorled Cladostephus.

Frond from four inches to nearly a foot high, irregularly branched; branches slender, covered throughout with whorls of jointed, mostly forked branchlets. Fructification formed in winter, when most of the whorled branchlets fall off, and are succeeded by accessory branchlets which bear the spores.

This plant is very common on our coasts, and is widely distributed elsewhere. It is frequently infested with parasitic algæ, especially with *Jania rubens*.

Cladostephus spongiosus. Spongy Cladostephus.

Fronds three to four inches high, irregularly branched; branches thick, densely covered with jointed, mostly simple, branchlets. Fructification on short accessory branchlets, produced in winter in the same manner as in the last species.

Dr. Harvey, in his 'Nereis Boreali-Americana,' expresses a doubt whether these two species are distinct. They differ in size and general appearance, and in the arrangement of the branchlets; but their technical characters are not constant.

Genus XXXIII. SPHACELARIA.

Frond branched, jointed, rigid, the tips of the branches distended into a membrane containing a granular mass called a *propagulum*. Fructification, egg-shaped spores, with a pellucid margin, or *perispore*, affixed to the branches, and propagula.—SPHACELARIA, from the Greek *sphakelos*, a gangrene, in allusion to the withered appearance of the tips of the branches.

This genus comprises several species, mostly natives of the shores of Europe; some of them extend to the Mediterranean, and even as far south as the Cape of Good Hope, while others are found in the Baltic and on the shores of Greenland.

Sphacelaria filicina. Fern-like Sphacelaria.

Frond shaggy at the base, slender, irregularly branched, from an inch to four inches high. Spores solitary, formed in the axils of the branchlets.

When growing on rocks, etc., in shallow water, this species is more robust and rigid than when parasitic on algæ at greater depths. In the 'Phycologia Britannica' the latter form is described and figured as a distinct species under the name *S. sertularia*; but Dr. Harvey evidently leans to the opinion that it is only a variety, and instances several analogous variations in other species. Subsequent writers have confirmed this opinion, which I am the more ready to adopt as the species of this genus are very prone to put on different appearances, and are in consequence very puzzling to young collectors.

Sphacelaria scoparia. Broom-like Sphacelaria.

Frond coarse, very rigid, from two to four inches high, of a dark-brown colour, the lower part of the stem covered with woolly fibres. Spores in bunches in the axils of the branchlets.

This species has also two distinct states dependent on the season of the year. In summer it is robust and rigid, and its stems are densely set with tufts of branchlets. In winter it becomes slender and delicate, and only single branchlets occupy the places of the tufts. There are, however, intermediate forms, which indubitably connect these two extremes.

Sphacelaria plumosa. Feathery Sphacelaria.

Frond naked at the base, irregularly branched, from two

to six inches long, branches comb-like; branchlets opposite, simple, very long, and closely set. Fructification unknown.

This species is rare, and grows but very little above low-water mark. It is very rigid, and has almost the appearance of a Sertularia.

Sphacelaria cirrhosa. Hair-like Sphacelaria.

Fronds growing in tufts on small sea-weeds, from a quarter of an inch to two inches long, slender, naked at the base, jointed throughout, more or less branched, branches closely set, either opposite or alternate. Fructification, globular spores attached to the branches, some on short stalks, others sessile.

Dr. Harvey very truly describes this species as "a very common and very variable plant, which puts on several distinct-looking forms, according to the locality in which it may grow."

Sphacelaria fusca. Brown Sphacelaria.

Fronds growing in dense tufts on rocks, very slender, one to two inches high, irregularly branched; branches erect, of the same thickness throughout; branchlets at distant intervals, minute, with star-like tips. Spores said to be "globose, scattered, sometimes stalked."

A very rare plant, the recorded habitats of which are in Wales and on the south-west coast of England.

Sphacelaria radicans. Rooting Sphacelaria.

Fronds hair-like, from half an inch to an inch long, erect or spreading, growing on rocks, in tufts which are combined into patches of various sizes; branches few, erect, straight, jointed, without branchlets, but with root-like fibres at the lower part. Fructification, globular spores, without stalks, either solitary or in clusters.

This species is also rare. Bantry and Waterford, in Ireland; Orkney and Appin, in Scotland; and Torbay, Ilfracombe, the Land's End, and Mount's Bay, in England, are the only recorded habitats.

Sphacelaria racemosa.   Clustered Sphacelaria.

"An inch in height, tufted, olivaceous, somewhat rigid, the fronds dichotomous; articulations equal in length and breadth; capsules oval, racemose, pedunculate."—*Greville*.

The first specimen of this plant was found in the Frith of Forth about 1821, by my late kind friend, Sir John Richardson, whose recent death has deprived more than one department of science of a member whose place will not be easily supplied. The spot has been since repeatedly searched, but neither there nor, until recently, elsewhere were any additional specimens discovered. Our knowledge of the species was, therefore, confined entirely to one set of plants; but as these were fortunately in full fruit, the title of the plant to rank as a species could be as definitely determined as though a large number of specimens had been obtained. The plant has been subsequently discovered in moderate abundance in the Frith of Clyde, by Mr. Roger Hennedy.

Genus XXXIV.   ECTOCARPUS.

Frond jointed, much branched, flaccid. Fructification scattered spores of various shapes, and pod-like, transversely-striped granules, *propagula*, formed of the whole, or of a part of a branchlet. The spores are very rarely to be found.—ECTOCARPUS, from the Greek *ectos*, external, and *carpos*, fruit.

This genus contains a large number of species, which are mostly natives of temperate climates, and are very abundant on the coasts of England and France. They are not at first sight attractive, their colour being dull and their substance slimy. They are, moreover, less easy to name than many other sea-weeds, as their specific characters are neither constant nor well-defined. Under the microscope, however, where the delicate texture and beautiful and varied forms of their jointed branches are revealed, they appear to much greater advantage. Dr. Harvey considers the pod-like body, or *propagulum*, to be the most reliable and easily-determined character, and it is, therefore, very important to obtain specimens on which this organ is developed.

Ectocarpus siliculosus. Podded Ectocarpus.

Tufts parasitical on various large Sea-weeds, pale-olive green; threads very slender, and much branched. Pods lance-shaped, usually on short, and more rarely on long stalks.

Ectocarpus amphibius. Amphibious Ectocarpus.

Tufts short, soft, of pale-olive colour; threads very slender, forked. Pods linear, scarcely distinguishable from the ordinary branchlets.

This species grows in ditches of brackish water, and Dr. Harvey suggests that it may be only a variety of *E. siliculosus*, altered by the circumstances of its growth.

Ectocarpus fenestratus. Windowed Ectocarpus.

Tufts small, pale green. Threads very slender, not much branched. Pods on short stalks, oblong, when ripe dark-coloured and chequered, like the lattice of a window.

Ectocarpus fasciculatus. Fasciculate Ectocarpus.

Tufts dense, olive-coloured. Main threads not much divided; branchlets in twig-like bunches. Pods without stalks, awl-shaped, set close together.

Ectocarpus Hincksiæ. Miss Hincks's Ectocarpus.

Tufts parasitical on *Laminaria bulbosa*, dark olive. Threads distantly branched, somewhat rigid; branches furnished on one side with curved, comb-like branchlets. Pods conical, without stems, forming the teeth of the comb-like branchlets.

This species is unhappily as rare as it is interesting and beautiful.

Ectocarpus tomentosus. Woolly Ectocarpus.

Tufts parasitical on *Fuci* and other large sea-weeds, commonly of a yellow-brown colour. Threads slender, irregularly branched, matted together into a spongy, cord-like mass of considerable length. Pods linear-oblong, on short stalks.

This species is very common, and its characters are better marked than those of any other British *Ectocarpus*. It usually attaches its matted masses of threads to the tips of the fronds of some *Fucus* or *Himanthalia*, and either floats lazily on the surface of a tide-pool, or hangs in ugly streaks among the rocks.

Ectocarpus crinitus. Hairy Ectocarpus.

Tufts growing on muddy sea-shores, from two to six inches long, spreading over the mud when the tide is down. Threads hair-like, very slender, and but little branched. Pods globular, without stalks.

The recorded localities for this rare species are Appin, Scotland, and Watermouth, near Ilfracombe.

Ectocarpus pusillus. The small Ectocarpus.

Tufts parasitical on small sea-weeds, having the appearance of "pale-brown wool." Threads slender, sparingly and irregularly branched, interwoven together. Pods roundish, on very short stalks, abundant.

This species, too, is rare. Hitherto it has only been found on the coasts of Devon and Cornwall. It probably exists in other localities, but has been overlooked in consequence of its small size.

Ectocarpus distortus. Distorted Ectocarpus.

Tufts parasitical on the leaves of *Zostera marina*, about six inches long, composed of numerous threads closely interwoven. Threads much and irregularly branched in a spreading, zig-zag manner. Pods egg-shaped, without stalks.

This is another rarity, not likely, I fear, to be found by many of my readers.

Ectocarpus Landsburgii. Landsborough's Ectocarpus.

Tufts small, intricate, growing in deep water. Threads tough, much branched in a zig-zag manner, and covered with short, spine-like branchlets.

This plant is very like *E. distortus,* but is of a tougher texture, and grows in deeper water. It is rare.

Ectocarpus litoralis. Littoral Ectocarpus.

Tufts parasitical on large sea-weeds, or growing on submerged wood or mud, from six inches to a foot long, matted, when young olive-green, becoming brown with age. Threads coarse, much branched. Pods in the form of linear swellings in the substance of the branches.

This is one of the most abundant and least attractive of British sea-weeds. It attaches itself to almost anything that comes in its way, and is not at all particular as to the depth of water in which it grows. It does not even confine itself exclusively to the sea, but may be found in estuaries and tidal rivers, sometimes far above the region of pure salt-water.

Ectocarpus longifructus. Long-fruited Ectocarpus.

Tufts parasitical on large sea-weeds, six or eight inches long, much branched. Threads coarse, with numerous, mostly opposite branches. Pods very long, tapering from the base to the apex, growing at the tips of the branchlets.

It is doubtful whether this plant be really distinct from *E. litoralis.* Dr. Harvey has figured it in the 'Phycologia Britannica,' but he states that he did so on the authority of a single specimen, and points out that the only differences between the two plants are the greater luxuriance and taper, terminal pods of *E. longifructus.* Among some specimens of undoubted *E. litoralis,* which I collected near Ilfracombe, were a few plants which, while possessing all the other characters of that species, had terminal pods similar to those ascribed to *E. longifructus*. I feel very little doubt that these were only a variety of *E. litoralis*, and I am inclined to believe that the two plants are but two different forms of the same species.

Ectocarpus granulosus. Granular Ectocarpus.

Tufts growing on rocks, corallines, or sea-weeds, in pools between the tide-marks, four to eight inches long. Threads with a principal stem, from which spring side-branches of unequal length, furnished with opposite branchlets, the whole having a feathery appearance. Pods elliptical, without stalks, scattered freely over the branchlets.

This species is abundant, and its characters are not difficult to make out, provided the specimens be in fruit.

Ectocarpus sphærophorus. Warted Ectocarpus.

Tufts parasitical on small sea-weeds, an inch to three inches high, very dense. Threads somewhat matted, having many times divided branches. Pods spherical, with a pellucid margin, seated on the branches, either opposite to one another, or to a branchlet.

This species, although found in several localities, is nowhere abundant. It generally grows on *Ptilota elegans*, and less frequently on *Cladophora rupestris*.

Ectocarpus brachiatus. Cross-branched Ectocarpus.

Tufts two to four inches high, feathery. Threads excessively branched; branches and branchlets opposite, main stems tangled. Pods immersed in the joints of the branches.

This is a very beautiful and rare species, and grows in brackish water, as well as in the sea.

Ectocarpus Mertensii. Mertens's Ectoca_rpus.
Tufts growing on mud-covered rocks and stones near low-water mark, dense but not matted, from two to six inches or more long. Threads branched; branches and branchlets opposite, of unequal lengths, intermixed. Pods imbedded in the branchlets.

A very pretty and well-marked species, occurring in several localities, but which must, nevertheless, be considered rare. I have gathered it at Worthing, and I have some splendid specimens which were collected at Plymouth, by Mr. John Gatcombe.

Genus XXXV. MYRIOTRICHIA.

Fronds jointed, hair-like, limp, beset with simple, spine-like branchlets, which are covered with pellucid fibres. Fructification, elliptical spores without stalks, each enveloped in a transparent membrane.—MYRIOTRICHIA, from the Greek *murios*, a thousand, and *thrix*, a hair.

All the species of this genus are parasitical, and occur frequently on our shores. They have neither bright colours nor beautiful structure to recommend them to notice; but they should, nevertheless, find their appointed place in every collection.

Myriotrichia clavæformis. Club-shaped Myriotrichia.
Fronds growing in tufts on *Chorda lomentaria*, about half an inch long, closely beset with branchlets, which in-

crease in length towards the tip of the frond, giving it the shape of a fox's brush. Spores without stalks, attached to the main thread.

Myriotrichia filiformis. Thread-like Myriotrichia.

Fronds growing on *Chorda lomentaria*, often in company with the last species, an inch or more in length, beset at irregular intervals with oblong clusters of short branchlets, which give them the appearance of being knotted, both the main threads and the branchlets clothed with long, jointed fibres. Spores without stalks, scattered along the main thread.

The branchlets of this species are short, and all of the the same length, and are set on in clusters at intervals; those of *M. clavæformis* are longer, increasing in length towards the tip of the frond, which is evenly covered with them throughout. With these exceptions, there is but little difference between the two species, and they frequently grow together on the same plant.

RED SEA-WEEDS.—RHODOSPERMEÆ.

Order VII. RHODOMELACEÆ.

Red or Brown-red or Purple Sea-weeds, with a leafy, or thread-like, or jointed frond composed of many-sided cells. Fructification of two kinds on different plants :—1. Pear-shaped spores in external ovate or urn-shaped conceptacles. 2. Tetraspores lodged either in distorted branchlets or in receptacles, called stichidia.

Genus XXXVI. ODONTHALIA.

Frond flat, thickish, distichously branched, indistinctly midribbed, alternately toothed at the margin. Spores in marginal, stalked, egg-shaped, wide-mouthed conceptacles; tetraspores tripartite, arranged in double rows in marginal, stalked, lance-shaped stichidia.—ODONTHALIA, from the Greek *odous*, a tooth, and *thalos*, a branch.

All the species of this genus are natives of cold latitudes; one is common to the northern shores of Europe and America, and the remainder have only been found on the north-east coasts of Asia.

Odonthalia dentata. Toothed Odonthalia.

Fronds rising from a hard disc, tufted, from three to twelve inches high, much branched, having a midrib in the lower part, and becoming flat and membranous above; branches springing from the axils of the teeth of the main stem, tapering at the base, deeply pinnatifid. Fructification of both kinds on the margin of the frond or in the axils of the teeth.

This plant, when growing, is of a deep blood-red colour; but becomes dark purple, almost black in drying. In this country it is confined entirely to the shores of Scotland and of the northern counties of England and Ireland, and within these limits occurs pretty frequently, being most abundant in the higher latitudes. I have gathered it on the coast of Yorkshire, near Whitby, and so far as I am aware, this is the most southern habitat that has been recorded.

Genus XXXVII. CHONDRIA.

Fronds thread-like, cartilaginous, pinnately divided, coated with small, many-angled, irregularly placed cells; axis jointed, many-siphoned; branchlets club-shaped, tapering abruptly at the base, obtuse, or nearly so, at the tip, transversely striped. Fructification:—1. Tufts of pear-shaped spores on simple threads radiating from a basal placenta contained, within a cellular pericarp, in ovate, perforate conceptacles, borne, either on stalks or sessile, on the branchlets. 2. Tripartite tetraspores crowded irregularly in the club-shaped branchlets.—CHONDRIA, from the Greek *chondrus*, cartilage.

This is a large genus, but has only two representatives among our native Sea-weeds. The species that compose it were formerly arranged among the *Laurenciæ*, which they resemble in external appearance, but from which they differ in structure. The many-siphoned, jointed axis of the stem is the character that is principally relied on to separate the two genera.

Chondria dasyphylla. The thick-leaved Chondria.

Frond robust, elongate, cylindrical, from six inches to a foot long; stem generally undivided, set with side branches which are either simple or furnished with a second or third series; branchlets about half an inch long, club-shaped, blunt, much constricted at the base. Spores in ovate, stalkless conceptacles; tetraspores immersed in the branchlets.

This is a very widely distributed species, and is found on several parts of our coast. It grows on shells or algæ, in shallow, sandy pools between the tide-marks. It is annual and in perfection in summer, when it should be of a dark purple colour; but this is frequently changed to a pale pink or even yellow by the action of the sun.

Chondria dasyphylla, var. squarrosa.

Tufts intricate, very dense. Fronds irregularly branched, robust, of a crisp texture; branches long, with a lanceolate outline; branchlets short, densely tufted. The whole plant of a dark brownish-purple colour.

I have found large quantities of this form in Jersey late in autumn. It is a very handsome plant, and very distinct in external appearace from the normal state of the species.

Chondria tenuissima. The slender Chondria.

Frond slender, irregularly divided, from a few inches to a foot long; branches long, rod-like, clothed with very slender, bristle-like, acute branchlets, which taper from the middle towards either extremity. Spores in ovate, nearly stalkless conceptacles, which are very numerous on the sides of the branchlets; tetraspores globose, scattered.

This species resembles *C. dasyphylla* in general appearance; but may be readily distinguished by its more slender habit, long, simple branches, and bristle-like, acute branchlets. Both these species were formerly included in the genus *Laurencia*, and are so placed in Dr. Harvey's 'Phycologia;' but in his later works he has removed them to the position which they occupy here. In St. Clement's Bay, Jersey, I have seen acres of this pretty little plant growing in the shallow, sandy pools near high-water mark; but all so discoloured that not a single purple or even pink specimen was obtainable. Indeed, had I not been well acquainted with the habit of the plant, I should have said it could not be a Red weed at all; but must belong to the class of Olives.

Genus XXXVIII. RHODOMELA.

Frond cartilaginous, composed of numerous, densely packed cells, cylindrical or slightly flattened, without joints, irregularly branched; branchlets slender. Spores pear-shaped, arranged in tufts in ovate conceptacles with or without stalks; tetraspores tripartite or cruciate, in single rows, imbedded in the tips of the branchlets.—RHODOMELA, from the Greek *rhodeos*, red, and *melas*, black.

The species which compose this genus are distributed over the colder regions of both the north and south temperate zones. They are closely allied to the genera *Rytiphlœa* and *Polysiphonia;* but differ from them in being of a more dense substance and in the absence of all appearance of joints.

Rhodomela lycopodioides. Lycopodium Rhodomela.

Fronds cylindrical, tapering upwards, from six inches to

two feet high, divided into long, simple branches, which are thickly clothed with slender, much divided, irregular branchlets. Spores in ovate conceptacles on the summer branchlets; tetraspores in the substance of the tips of the winter branchlets.

The summer and winter states of this plant are so different that they may be easily mistaken for two species. In the former the side branches are from one to three inches, or more, long, and furnished with numerous, slender, much divided branchlets; in the latter they are simple, or only slightly divided, and seldom more than half an inch long, The colour of the plant when growing is a dark red, which changes to black in drying. This property is common to all the species of the genus, and hence the appropriate name of redblack.

Rhodomela subfusca. Brown Rhodomela.

Fronds tufted, much branched, from six inches to a foot long, cartilaginous, rigid; branches irregularly divided, the first series alternate, the second twice pinnate, the third furnished, below with simple, alternate, rather distant, awl-shaped branchlets, above with more divided branchlets, which are densely crowded at the ends of the branches. Spores in ovate, very short-stalked conceptacles; tetraspores arranged in pairs, or singly in the awl-shaped, terminal branchlets.

This species is common all round our coasts. Like the last, it has two very distinct states. In summer it appears in all the perfection of parts described above. In winter only a few almost naked stems are to be found; but as these frequently bear accessory stichidia containing tetraspores, they should be as carefully preserved a their more luxuriant brethren.

Genus XXXIX. BOSTRYCHIA.

Frond of a dull purple colour, thread-like, inarticulate, composed of a tubular, jointed axis, surrounded by oblong coloured cells, which become shorter towards the circumference, those at the surface being square, and giving the plant the appearance of being dotted. Spores pear-shaped, arranged in tufts in ovate, terminal conceptacles; tetraspores tripartite, in double rows, in spindle-shaped, terminal stichidia.—BOSTRYCHIA, from the Greek *bostrychos*, a curl.

This is a rather extensive genus of small sea-weeds, all of a dull purple colour, and growing usually near high-water mark, or in places which are only occasionally submerged. They have no objection to brackish or even perfectly fresh water, and are sometimes found in streams at a considerable distance from the sea. These peculiarities are the more remarkable, as most of the species of the allied genera grow in depths where they are rarely exposed by the receding tide, and where they are far removed from all chance of contact with fresh water. The species are widely distributed in warm and temperate latitudes, several being abundant at the mouths of the rivers running into the Gulf of Florida, and on the adjacent coasts of North and South America; one occurring in Kerguelen's Land, and others on the Atlantic shores of Europe.

Bostrychia scorpioides. Scorpion-like Bostrychia.

Fronds two to four inches high, growing in matted tufts,

branched at intervals; the branches three or four times pinnated, beset with spreading or reflexed branchlets, which are about half an inch long, and much more slender than the main stems. The tips of both the branches and branchlets are strongly incurved. Neither spores nor tetraspores are usually found on British specimens.

This species is the only representative of the genus in our flora,—I was going to write marine flora, but that would not have been strictly correct, for it is commonly found among the roots of flowering plants at the estuaries of rivers, and in salt ditches and marshes.

Genus XL. RYTIPHLŒA.

Frond cylindrical or flattened, shrub-like, striped crosswise, composed of a jointed axis of large tubular, elongated cells of equal length, surrounded by a periphery of one or more series of irregularly-shaped, small, coloured cells. Spores pear-shaped, arranged in tufts in ovate conceptacles with or without stalks; tetraspores in terminal, spindle-shaped branchlets, or in single or double rows, in stichidia.—RYTIPHLŒA, from the Greek *rutis*, a wrinkle, and *phloios*, bark.

The characters of this genus are very similar to those of some species of *Polysiphonia*, the most easily defined difference between the two genera being, that in *Rytiphlœa* the frond is composed of nearly equal parts of axis and periphery, and consequently the joints of the former appear at the surface only as stripes, while in *Polysiphonia* nearly the whole bulk of the frond consists of jointed tubes, similar to those in the axis of *Rytiphlœa*; and either there is no periphery, or it is a mere bark, and the articulated structure of the plant is, therefore, clearly visible. The species are not numerous; but the genus is well represented on our coasts, where three kinds are abundant, and a fourth is occasionally found. The remainder are natives of tropical, or almost tropical regions.

Rytiphlœa pinastroides. Pine-like Rytiphlœa.

Fronds from a few inches to about a foot high, cylindrical, irregularly and densely branched, shrub-like; the main branches alternate, their lower parts clothed with short, awl-shaped branchlets; the secondary branches elongate, spreading or recurved, curled inwards at the tip, their upper side furnished with two rows of straight or hooked branchlets, arranged generally in pairs, and giving the branch a comb-like appearance. Spores in ovate, stalked conceptacles on the inner sides of the upper branchlets; tetraspores tripartite, in double rows, in lance-shaped, stalked stichidia.

This plant is in perfection in winter and in early spring, and is abundant in many localities on the south coast and in the Channel Islands. It grows on submarine rocks, near low-water mark, and is, therefore, generally beyond the reach of the collector. Numerous specimens are, however, thrown up during every period of rough weather. When fresh, it is of a dull red colour, which changes to black in drying. It will not adhere to paper.

Rytiphlœa complanata. Flat Rytiphlœa.

Fronds from two to four inches high, shrub-like, flat, alternately branched, of a brown-red colour, marked throughout with curved, transverse stripes; the lower part of the stem naked, or sparingly furnished with short branchlets;

the branches thickly set with alternate branchlets, which are short and simple at the base of the branch, and become gradually longer and more divided as they ascend. Spores in roundish, stalkless conceptacles; tetraspores in the branchlets, near the top of the frond. I am, however, unable to find any record of either spores or tetraspores on British specimens.

This is one of the most rare of our native Sea-weeds, and has only been hitherto found at Bantry Bay and one or two other localities in the south of Ireland, and on the coasts of Devon and Cornwall. It is a summer plant, and grows in shallow tide-pools. It requires to be soaked for some hours in fresh water before being laid down, or it will neither preserve its colour nor stick to paper. I am indebted to my friend Mr. John Gatcombe for a specimen of this rare plant, collected by him at Plymouth.

Rytiphlœa thuyoides. Arbor-vitæ-like Rytiphlœa.

Fronds from three to six inches high, shrub-like, rigid, growing in tufts, of a dark brownish-purple colour, striped crosswise; stems erect, cylindrical, rising from a mass of creeping, fibrous roots, the lower part simple, the upper part irregularly divided into alternate branches, the whole covered with short, spine-like branchlets, those below simple, those above forkedly branched. Spores pear-shaped, arranged in tufts, in egg-shaped, stalkless conceptacles, which are usually very abundant; tetraspores tripartite, in blunt, distorted branchlets.

This species is pretty generally distributed round our coasts, and is especially plentiful on the western shores of Ireland. It grows on rocks, or on small sea-weeds in tide-pools, and varies in size according to the depth of the water. In places occasionally exposed to the air it is stunted. Near low-tide mark it is more luxuriant. It is perennial, and in perfection in summer.

Rytiphlœa fruticulosa. Shrubby Rytiphlœa.

Fronds from four to six inches long, irregularly branched from the base upwards, striped crosswise throughout, of a dark purple colour, which changes to a more or less yellow-green when the plant is exposed to the rays of the sun. Stems cylindrical, tapering upwards, intertwined; branches forked, spreading, much and irregularly divided above; branchlets alternate, spreading, thickly studded over the whole frond, those near the base not more than a quarter of an inch long, those higher up longer, and furnished with awl-shaped secondary branchlets. Spores in numerous, stalkless, almost round conceptacles; tetraspores tripartite, in distorted branchlets.

This species, like the last, grows on rocks or sea-weeds, in tide-pools, where its spreading, spurred branches become entangled with each other, or with neighbouring sea-weeds into masses of considerable size. It is perennial, and may be found during summer in most localities on our southern and western shores. Some of the praises which poets have bestowed on the loveliness of a rose bedecked with dew may fairly be claimed for this little plant, which, when fresh drawn from the sea, presents the same kind of beauty under another aspect. Each of its terminal branchlets brings with it from its briny home a tiny, sparkling drop of water which clings to it, as though in a last embrace, for several seconds in the upper air. These bright beads of water give to the whole plant a transient gleam of radiance that is very charming.

Genus XLI. POLYSIPHONIA.

Fronds thread-like, rarely a little flattened, jointed, or with a jointed axis, composed of elongated tubular cells, arranged round a central cell. Spores pear-shaped, in ovate or urn-shaped conceptacles; tetraspores (of British species) tripartite, imbedded in distorted, terminal branchlets.—POLYSIPHONIA, from the Greek *polus*, many, and *siphon*, a tube.

This genus contains, according to various writers, between two and three hundred species, which are distributed in almost every latitude between the poles and the equator, and vary much in size and in external appearance. Dr. Harvey describes it so admirably in his 'Nereis Boreali-Americana,' that I shall quote his words. He says:—"Some species are two to three feet in length, others not more than as many tenths of an inch; some dichotomous, others pinnated—some distichous and fern-like, others with a bushy or arborescent character; some of cobweb delicacy, lubricous, and excessively flaccid, soon decomposing, others robust, rigid or tough, of strong enduring substance; some of a brilliant rosy-red or crimson, others (and the greater number) varying through all the graver shades of red-brown, brown, and purple; some inhabiting the deep sea, others occurring only near high-water mark or far up the estuaries of tidal rivers. Plants of such varied aspect and habit could not have been brought together by the universal consent of botanists, among whom there has never been much difference of opinion respecting the just limits of this genus, if they had not

H

some obvious bond of union in an essential, easily seen, and important common character. This is found in the structure of the stem in the articulated species, and of the axis of the stem in species which appear to be partially inarticulate; the dissepiments being hid by the growth of a thin or thick layer of epidermal cells round the stem or branches."

The tubes, or siphons, which constitute the character of this genus, serve also to distinguish the various species, and to enable the student to arrange these in the most natural subdivisions. Recent writers on the subject are not, however, agreed as to the best mode of doing this; for while some have merely grouped together those species which possess common characters, others have divided the genus into two or more subgenera; and Dr. J. E. Gray, in his 'Handbook of British Water-weeds,' has gone so far as to separate even those species which occur on our coasts into four distinct genera. The difference between these various modes is not really so important as might be supposed. It arises in part probably from a desire not to disturb existing genera more than is absolutely necessary, and in part from a variety of opinion as to whether certain characters should be considered generic, or only specific, or subgeneric. Thus we find Dr. Gray's genus *Polysiphonia* corresponds almost exactly with the subgenus of the same name in Dr. Harvey's ' Nereis Boreali-Americana;' and so also the genus *Oligosiphonia*, of the former author, with the subgenus of the same name of the latter. Leaving higher authorities to determine which is the best arrangement, I shall merely group together the allied species, and thereby avoid the use of names not hitherto known to collectors of British Sea-weeds.

The more delicate species of this genus require to be laid down in salt-water, or they will lose their colour and stain the paper on which they are dried. At certain periods of their growth, particularly when nearly mature and in fruit, they decay so quickly that it is necessary to lay them out almost immediately after being gathered.

Subdivision 1.—*Primary tubes more than four.*

Polysiphonia Brodiæi. Brodie's Polysiphonia.

Fronds from six to twelve inches long; stem composed of an axis of seven primary and seven secondary tubes, surrounded by a thick layer of smaller cells, which form a bark, and hide the joints of the axis; branches alternate, furnished with numerous short tufts of delicate jointed branchlets; articulations rather longer than broad. Spores in egg-shaped, short-stalked conceptacles; tetraspores in swollen branchlets.

This large and handsome species grows on rocks and corallines, near low-water mark. It is moderately abundant in many localities, and thrives equally in the comparatively cold climate of Scotland and in the warmer latitude of the Channel Islands. It is annual, and in perfection in summer.

Polysiphonia nigrescens. Blackish Polysiphonia.

Fronds shrub-like, from three to twelve inches or more long, of very variable thickness, growing several from the same base; branches alternate, pinnate; branchlets elongated, awl-shaped, alternate; tubes flat, about twenty in number, placed round a large central cavity; articulations very short. Spores in broadly-ovate, stalkless conceptacles; tetraspores near the tips of the branchlets.

Specimens of this species vary very much, according

to the season in which they may be gathered. In spring and summer, the fronds have their full complement of branchlets, and are in consequence light and feathery. In autumn and winter, the more slender branchlets are wanting, and the plant has a coarse, worn, and scrubby appearance. This is not a difficult species to determine. The numerous tubes and short joints of its stem are characters whereby it may be readily distinguished. It is perennial, and may be found at any season on almost every part of our coast.

Polysiphonia atro-rubescens. Dark-red Polysiphonia.

Fronds erect, rigid, densely tufted, from three to twelve inches long; main branches forked, long, tapering; branchlets alternate, short, spine-like, fastigiate, more or less forkedly divided; tubes, about twelve, arranged spirally round a small central cavity; articulations, of the lower part of the frond, about three times as long as broad, those of the upper part about once and a half. Spores in broad, wide-mouthed, short-stalked conceptacles; tetraspores in spindle-shaped branchlets.

This species grows near low-water mark, and is moderately abundant in many localities. It is annual, and in perfection in summer and autumn. The spiral arrangement of its tubes, taken in conjunction with its other characteristics, will enable the young student to distinguish it without much difficulty.

Polysiphonia subulifera. Awl-bearing Polysiphonia.

Fronds tufted, flaccid, from five to eight inches long, about as thick as a bristle; main branches irregularly forked below and divided alternately above; lesser branches long, and but little divided; branchlets short, spine-like, awl-shaped, scattered over the whole plant; tubes in the stem thirteen; articulations varying, in different specimens and in different parts of the same specimen, from once to thrice as long as broad, visible in all parts of the frond. Spores unknown; tetraspores in forked branchlets.

In consequence probably of being a native of deep water this species has been seldom found, and is considered rare. Externally it bears some resemblance to *Rytiphlœa fruticulosa*, but may be easily known by its more slender habit, and the distinct joints of its stem and branchlets. It is annual, and grows during summer, generally on Nullipore banks, in from four to ten fathoms of water.

Polysiphonia obscura. Hidden Polysiphonia.

Fronds densely matted together, spreading over rocks or the roots of large sea-weeds, throwing out root-like processes downwards, and erect undivided branches about half an inch long upwards; tubes about thirteen; articulations not quite as long as broad, visible in all parts of the frond. Tetraspores in thickened pod-like branchlets.

This plant may well be called the *hidden Polysiphonia*; for it is so tiny that only a diligent seeker has a chance of finding it. I have no doubt that it grows in many localities where it has been hitherto overlooked.

Polysiphonia parasitica. Parasitic Polysiphonia.

Fronds slender, rigid, feathery, alternately branched, from one to three inches long, growing several together from the same base; branches tripinnate; branchlets closely set, alternate, awl-shaped; tubes about eight, arranged round a small cavity; articulations about as long as broad. Spores in ovate, stalked conceptacles; tetraspores in swollen branchlets.

This rare and very beautiful species is pretty generally distributed round our coasts. It is only to be found at and beyond extreme low-water mark, where it is parasitic on the *Melobesiæ* and *Corallinæ* which grow on the perpendicular sides of ledges of rock. None but floating specimens can, therefore, be obtained except at the lowest spring tides, or by dredging. It is to this circumstance, probably, that its reputed rarity is partly due.

Polysiphonia variegata. Variegated Polysiphonia.

Fronds growing in dense tufts, from four to ten inches long, about as thick as a bristle at the base, gradually tapering upwards to the size of a hair, much branched; tubes six, or rarely seven, arranged round a minute central cavity; articulations, in the lower part of the stems, shorter than they are broad, in the branches and branchlets, varying in length from once to twice their breadth, with three broad tubes distinctly visible in all parts of the frond. Spore-conceptacles on short stalks on the smaller branches and branchlets; tetraspores small, in slightly swollen branchlets.

This species grows on mud-covered rocks, floating timber, *Zostera*, etc. In the few localities where it has been hitherto found, chiefly on the coasts of Devon and Dorset, it is both luxuriant and abundant. Its distinctive characters are the six tubes of its stem and its purple colour. It is annual, and in perfection in summer.

Polysiphonia furcellata. Forked Polysiphonia.

Fronds growing in dense, tangled tufts, five or six inches long, about as thick as a bristle below, gradually tapering upwards, much divided into forked branches, which spring from spreading, rounded axils; tubes about eight; articulations in the lower parts of the fronds, from three to five times as long as broad, in the upper, of about equal length and breadth. Spores in somewhat globular, stalkless conceptacles; tetraspores in distorted branchlets.

This species grows on rocks in deep water, and is very rare. When fresh it is of a bright brick-red colour. It is a summer annual.

SUBDIVISION 2.—*Primary tubes four.*

Polysiphonia urceolata. Pitcher-shaped Polysiphonia.

Fronds growing in tufts, connected by root-like fibres, erect, rigid, from three to nine inches long, much and irregularly branched; articulations, of the lower part of the stem, about as long as broad, of the branches, from three to five times as long as broad, and of the upper branchlets, shorter than broad. Spores in pitcher-shaped, stalked conceptacles; tetraspores in the upper part of the branchlets.

This elegant plant is abundant all round our coast. It grows, during summer, on rocks near low-water mark, and in deep water, on the stems of *Laminaria digitata*, the specimens found in the former situation being usually larger and more robust than those in the latter. The most obvious characters of the species are the short articulations of its stem, which are marked by two broad tubes, and its bright red colour. It is annual.

Polysiphonia formosa. The beautiful Polysiphonia.

Fronds densely tufted, from six to ten inches long, erect, very slender, divided into long, flexuose branches and branchlets, whose tips are sometimes fibrillose; articulations mostly many times longer than broad. Spores in pitcher-shaped, stalked conceptacles; tetraspores in single rows, in spindle-shaped branchlets.

Although by no means common, this species is distributed over the coast of the British Isles from Shetland to the south of England. It grows on rocks near low-water mark, chiefly in bays and estuaries, and is considered by some writers to be only a very slender form of *P. urceolata*, from which species it differs in the much greater delicacy of its fronds, and the longer articulations of its stems. It is annual, and grows in summer.

Polysiphonia fibrata. The fibred Polysiphonia.

Fronds growing in very dense tufts, from two to eight inches long, as thick as a bristle at the base, tapering upwards, erect, gelatinous, of a dark red colour; branches dichotomous; branchlets tipped with tufts of jointed fibres; articulations mostly several times longer than broad. Spores in ovate, wide-mouthed, stalked conceptacles; tetraspores small, in distorted branchlets.

This is another widely distributed species. It is annual, and grows, during summer and autumn, on rocks and algæ in tide-pools near low-water mark, and occasionally in more exposed places. The fibres at the tips of its branches, from which the name *fibrata* is taken, although not peculiar to it, for most *Polysiphoniæ* have them in certain stages of their growth, are more constant and abundant in this than in any other species except *P. fibrillosa*, and, taken in conjunction with other characters, afford the best means of determining its identity. In some specimens these fibres are the seat of oblong yellow bodies, called *antheridia*.

Polysiphonia pulvinata. Cushioned Polysiphonia.

Fronds spreading over rocks in dense roundish tufts about an inch high; branches springing at right angles from the creeping stem, erect, forked, flaccid, very slender; branchlets short, scattered, spreading, sometimes curved; articulations, in the main branches, about three times as long as broad, in the branchlets, about half the length of their breadth. Spores in comparatively large, narrow-mouthed, urn-shaped, stalked conceptacles; tetraspores in single rows, in the branchlets.

With the exception of *P. obscura*, this is the smallest of our native *Polysiphoniæ*. It is annual and grows on rocks between the tide-marks, in moderate abundance in many localities. Its small size, creeping habit, and general appearance render it an easy species to determine.

Polysiphonia Griffithsiana. Mrs. Griffiths's Polysiphonia.

Fronds about four inches long; stem simple; branches alternate, spreading, those nearest the base longer than those next above them, and thus decreasing gradually as they approach the apex of the frond; branchlets dichotomous, flaccid, slender; tubes, four primary and four secondary; articulations about once and a half as long as broad. Spores in ovate, stalkless conceptacles.

This very pretty plant was discovered at Torquay in 1837 by the distinguished algologist whose name it bears. It has since been found in one or two other localities, but must still be considered a great rarity. It grows parasitically on *Polyides rotundus* and other small algæ, between high- and low-water mark. It is annual, and to be found at the end of summer and in early autumn.

Polysiphonia spinulosa. Round-fruited Polysiphonia.

" Dark-red; branches divaricate, somewhat rigid, the ramuli short, straight, subulate, divaricate; articulations about equal in length and breadth, three-tubed. Tubercles (spore conceptacles) globose, sessile, excessively minute."— *Greville.*

So little is known of this plant, that it is difficult to determine whether it be a distinct species, or merely a variety. The only recorded specimens were found many years since by Captain Carmichael, at Appin. The above description is copied *verbatim* from Dr. Greville's 'Scottish Cryptogamic Flora.'

Polysiphonia Richardsonii. Sir John Richardson's Polysiphonia.

" Stems cartilaginous, setaceous; branches alternate, elongated, divaricate, beset in the upper part with very patent, straight, subdichotomous ramuli; articulations of the stem and branches two or three times longer than broad, irregularly veined; of the ramuli shorter. Capsules sessile, globose."—*Harvey, Phyc. Brit.*

This is another very doubtful species, the only known

specimens of which were gathered by the late Sir John Richardson, before his first Arctic expedition. Dr. Harvey, whose description of it I have copied *verbatim,* suggests that it may be a variety of *P. fibrillosa,* and mentions three or four other species that it more or less resembles.

Polysiphonia elongella. The divaricate Polysiphonia.

Fronds from two to five inches high, growing singly, or two or three together; stem, at the base, about as thick as a bristle, rigid, above, more slender and flexuose; branches irregularly forked, spreading, springing from wide axils; branchlets slender, tufted, and of a blood-red colour; articulations visible throughout the frond, about as long as broad, except those in the lower part of the stem, and at the tips of the branchlets, which are shorter. Spores in large, egg-shaped, stalked conceptacles; tetraspores immersed in the branchlets.

This is not a very common species, but has been found in several places, both on our northern and southern shores. It is biennial, and grows during the greater part of the year on rocks or small Sea-weeds, at and beyond low-water mark. It varies very much in appearance according to the season in which it may be gathered. Dr. Harvey writes, "The winter and summer aspects of a deciduous tree are not more different from each other than are specimens of this beautiful plant collected at opposite seasons."

Polysiphonia elongata. Lobster-horn Polysiphonia.

Fronds from four to twelve inches high, growing singly,

or two or three together; stem rigid, erect, as thick as small whipcord, or less, undivided and without branches below, much branched and bushy above; branches more or less irregularly divided, tapering abruptly at the base, and gradually at the tip; branchlets narrow-spindle-shaped, tapering towards either extremity, tipped with short fibrils, not numerous on young plants, falling off in winter, and reappearing in much greater luxuriance in the succeeding spring and summer; articulations of the stem and branches shorter than broad, partially hidden by the small cells which surround the tubes; those of the branchlets as long as, or longer than broad, and more distinctly visible. Spores in egg-shaped, stalkless conceptacles; tetraspores large, in swollen branchlets.

This species is perennial, and very common in most localities at all times of the year. It grows on rocks, shells, and pebbles, in tide-pools and in deep water. Specimens in the first season of their growth differ much from those in the second, and both change considerably as winter approaches. The slender branchlets fall off, like the leaves of a tree or land plant, and the stems and branches are left bare until the ensuing spring. It is from the appearance which it presents in this latter state that the English name of Lobster-horn is derived.

Polysiphonia violacea. The Violet Polysiphonia.

Fronds from a few inches to a foot long, of a brownish-red colour, becoming violet when dry; stem varying in thickness according to the size of the plant, irregularly divided, becoming broad and flat when laid out on paper; branches alternate, long, silky; branchlets very slender, tufted, tipped with fibres; articulations of the stem and large branches, hidden by a thick layer of irregular cells; those of the branchlets visible, the lower ones four times, the upper twice as long as broad. Spores in egg-shaped, stalked conceptacles; tetraspores roundish, in swollen branchlets.

This elegant plant is a native of many parts of our coast. It grows on rocks, stones, or sea-weeds, near low-water mark, attains perfection in the months of May and June, and is annual. It varies a good deal in size, and in the greater or less development of the branchlets, which in some specimens are very luxuriant. It is of a soft, silky texture, and adheres closely to paper.

Polysiphonia fibrillosa. The fibrillose Polysiphonia.

Fronds growing singly or in tufts, from four to ten inches long, of a pale-brown colour; stem thick, running distinctly through the frond, not at all, or only once or twice divided; branches alternate, spreading, thick; branchlets numerous, very slender, always tipped with jointed, forked fibres; articulations of the stem and branches, more or less hidden by narrow wavy cells; those of the lesser branches and branchlets visible, rather longer than broad. Spores in egg-shaped, stalkless conceptacles; tetraspores large, in the terminal branchlets.

This plant is annual, and common in most places during the summer. It grows in sunny pools between the tide-marks. In some respects it resembles *P. violacea*, but has shorter, less tufted branchlets and shorter articulations, and is of a paler colour than that species. It is also more constantly fibrillose.

Subdivision 3.—*Frond forked; tubes numerous, flattened, arranged round a large, jointed, central cavity containing bags of coloured endochrome.*

Polysiphonia fastigiata. Flat-topped Polysiphonia.

Fronds rigid, from one to two inches long, growing in globular tufts; stems as thick as horse-hair, much divided, forked; branches and branchlets forked, their axils acute, their tips awl-shaped; tubes about eighteen; articulations shorter than they are broad. Spores in egg-shaped, stalkless conceptacles; tetraspores immersed in the terminal branchlets.

This species grows parasitically on two or three kinds of *Fuci*, but most commonly on *F. nodosus*, which is very rarely free from it; wherever the larger plant is found, and that is almost everywhere, it is sure to be fringed with dark-brown, hair-like tufts of the smaller. Indeed the two are so nearly inseparable that they mutually impart a character, the one to the other. *P. fastigiata* is a perennial, and attains its greatest luxuriance of growth in summer and autumn. It possesses a very peculiar microscopic character, which may perhaps be overlooked if I do not specially describe it. I allude to the bags of endochrome which exist in the centre of the cavity round which the tubes are arranged. No other British species of *Polysiphonia* is furnished with these organs, on which Dr. Gray, in his 'Handbook of British Waterweeds,' has founded a new genus under the name of *Vertebralia*, in allusion to their being, as it were, the backbone of the plant. If a small section of the stem be crushed between two glass slides, and then examined under a microscope, these bags will be seen among the sections of the tubes, from which they may be readily distinguished by their greater breadth.

SUBDIVISION 4.—*Tubes seven, branches and branchlets beset with single-tubed, forked, jointed ramelli or leaves.*

Polysiphonia byssoides. The byssoid Polysiphonia.

Fronds from a few inches to a foot or more long; stem about as thick as a bristle, undivided throughout its entire length, becoming flat when laid out on paper; branches alternate, bearing one or two series of branchlets, all clothed with short, byssoid, jointed, ramelli or leaves; articulations varying much in length in different specimens: those of the stem and larger branches from twice to six times as long as broad; those of the smaller branches usually short. Spores in egg-shaped, stalkless conceptacles; tetraspores formed from the joints of the branchlets.

This is a large and handsome plant, and is very generally distributed round our coasts. It grows on stones, shells, and sea-weeds, at and beyond low-water mark. Possessing the characters of a *Polysiphonia*, in combination with the byssoid ramelli of a *Dasya*, this plant may be said to be the connecting link between the two genera. Dr. Gray has placed it in a separate, new genus, under the name *Dasyclonia*.

Genus XLII. DASYA.

Frond without visible joints, thread-like, or flat, branching, with a many-tubed, jointed axis, surrounded by a layer of cells; branches clothed with slender, single-tubed, jointed ramelli. Spores in egg-shaped, pointed conceptacles with or without stalks; tetraspores in lance-shaped stichidia formed on the ramelli.—DASYA, from the Greek *dasus*, hairy.

Feeling that nothing can excel the description which Dr. Harvey gives of this genus in his 'Nereis Boreali-Americana,' I again quote *verbatim* from that work. He writes, "A large and considerably diversified genus occurring in both hemispheres. As here understood, it is chiefly characterized by the confervoid, jointed ramelli, issuing from a compound polysiphonous, but mostly opaque, and outwardly inarticulate frond; and the lanceolate, pod-like receptacles of tetraspores, borne by the confervoid ramelli, out of whose branches they are formed. The ramelli are of the same structure as the articulated fibres which clothe the ends of the young branches in *Polysiphonia, Rhodomela,* etc., but in those genera they are mostly colourless, very fugacious, and have no connection with the tetrasporic fructification; in *Dasya*, on the contrary, they are persistent, containing coloured cells, and finally originating the tetrasporic fructification. In the former cases they accompany the early development of the branches only, in the latter they are characteristic of the species at all ages."

The British species of this genus, with the exception of *D. coccinea*, grow chiefly on our southern and western coasts, and are most abundant in the Channel Islands. In all the graces of colour, form, and texture, these exquisite little plants hold a high place among their compeers, and they also possess a virtue by no means common in pretty Sea-weeds—they are very docile under the operation of being laid out on paper, spreading themselves over its surface in the mode best adapted to display their beauties, and adhering to it closely when dry.

Dasya coccinea. Scarlet Dasya.

Root a small disc. Fronds from a few inches to nearly a foot high, growing singly, or in bunches; stem thick, clothed with short, hair-like fibres, unequally divided; branches alternate, feather-like; branchlets short, slender, much divided; articulations visible only in the smaller branches, very short. Spores in egg-shaped conceptacles; tetraspores in oblong, pointed, stalked stichidia.

This is a very abundant, well marked species, not at all liable to be confounded with any other British Seaweed. It grows on rocks, at and beyond low-water mark, and is most commonly found floating, or cast on shore by the tide, the finest specimens being obtainable at the end of summer, or in the early autumn. There is a small, depauperated variety, which grows on sand in deep water: it is without hair-like fibres, its branches are few and irregular, and its branchlets squarrose.

Dasya ocellata. The ocellate Dasya.

Root a small disc. Fronds from one to three inches high, growing in tufts; stems simple, or more or less forkedly divided, thickly covered throughout their length with long, slender, forked branchlets; articulations much longer than broad. Spores not hitherto observed in British specimens; tetraspores in lanceolate, short-stalked stichidia borne on the branchlets.

This species is annual, and grows during summer on mud-covered rocks, which are not exposed at low water, and are not, therefore, easy of access. This fact, and the very small size of the plant, combine to render it even more rare than the real scarcity of specimens warrants. The name *ocellata* has reference to the spots caused by the density of the branchlets at the tips of the fronds, which are supposed to resemble the eyes in the tail-feathers of a peacock.

Dasya arbuscula. The Shrub Dasya.

Root a small disc. Fronds from one to four inches high; stems much and irregularly branched, thickly set with short, forked branchlets; articulations about twice as long as broad. Spore-conceptacles roundish, with suddenly tapering, long, cylindrical necks, and short stalks; tetraspores in oblong, dagger-tipped stichidia; both the conceptacles and stichidia grow on the branchlets.

This species is annual, and to be found in summer. It is entitled to be considered one of the beauties of our marine flora, and is moderately abundant in the Channel Islands, and in some parts of Ireland and Scotland, but is rare in England. The most robust specimens grow on rocks at extreme low-water mark; those which are dredged in deep water are more slender, and less densely set with branchlets. The difference between extreme specimens of the two forms is so great as to induce the belief that they must be distinct species; but a careful examination of the intermediate varieties will prove that such is not the case. In this species, also, the branchlets become more dense at the tips of the frond, where they form little eye-like dots, or tufts.

Dasya venusta. The beautiful Dasya.

Root a small disc. Fronds two to four inches long, growing singly or several together; stem running through the frond, thick, cylindrical, when dry, flat; branches irregularly alternate, pinnate, or bipinnate; branchlets repeatedly forked, jointed, slender, tapering to extreme tenuity at the tips; articulations many times longer than broad. Sporeconceptacles inverted-pear-shaped, with short, broad necks, and short stalks; tetraspores in lanceolate, acuminate, short-stalked stichidia.

This plant well deserves the specific name of Beautiful, given to it by Dr. Harvey; for its bright rose-coloured fronds, with their graceful forms and delicate texture, do not yield the palm of loveliness to any member of this favoured family. It was first found in Jersey about twenty years since, and has subsequently proved to be moderately abundant on the shores of that island, but has not, so far as I am aware, been discovered in any other locality. This fact raises the question whether plants from the Channel Islands are entitled to be included in the list of British Sea-weeds. Dr. Harvey evidently considers that they are, for he has recorded Jersey habitats for a large proportion of the species described in his 'Phycologia,' and in my humble opinion he is quite right. Indeed I cannot bear to think of the many vacant spaces that there would be in my own collection were I to withdraw all the Channel Islands specimens. These islands have belonged, uninterruptedly, to the British Crown from the time of the Norman conquest to the present day, so that politically they are, without doubt, thoroughly English. Geographically, their claim is not so strong; but there is so little difference between the flora of our own country and that of the northern part of France, that this part of the question is not important. There are very few sea-weeds found in the Channel Islands which are not also natives of our own shores, and of the northern coasts of France. The difference is, that plants which are very rare with us, and are exclusively confined to certain favoured southern or western localities, are more abundant and luxuriant in the milder climate and congenial habitats to be found among the rocky bays of Jersey and her sister islands. *D. venusta* is annual, and is cast up from deep water during summer and autumn. It is one of the least difficult sea-weeds to lay out, and is perhaps the most exquisitely beautiful, at least of our native species, when laid out.

Dasya punicea. The purplish Dasya.

Fronds irregularly pinnately-branched, coated throughout with bark cells; the stem and lower branches naked, the upper clothed with very short, delicate, jointed branchlets. Spore-conceptacles broadly ovate, short-necked, stalkless; tetraspores in lanceolate, pointed stichidia.

The knowledge we possess of this species is derived from a very limited number of specimens collected in the Adriatic Sea, and in one or two localities on our southern shores. It was first discovered in this country by that indefatigable and accomplished algologist Mrs. Gray, of the British Museum, who found it at Bognor, in October, 1855. It has been subsequently picked up at Brighton by Mrs. Merrifield. I am not aware that any other British habitats have been recorded. I have had the advantage of examining Mrs. Gray's specimens.

Dasya Cattlowiæ. Cattlow's Dasya.

Frond about four inches high, with an irregular, rounded outline, flaccid; main stem robust, flat when dry, running distinctly through the frond, and furnished with branches quite down to its base; branches irregularly pinnate, opposite or alternate, long, slender, rounded at the tip, clothed with very long, delicate, many-times forked, jointed branchlets, which commence with a single fibre, but by repeated forkings become tufted. Fructification unknown.

This species was first found floating in St. Aubin's Bay, Jersey, by Miss Mary Cattlow, in 1858. The specimen was submitted to Dr. Harvey, who believed it to be distinct, and named it after the discoverer. He has not, that I am aware, published any description of it, and the only notices of the plant that I can find are a reference to it as "a form not yet described," in the appendix to Mrs. Gatty's 'British Sea-weeds,' and a brief description at the end of Dr. Gray's 'Handbook of Water-weeds.' I am indebted to Mrs. William Mauger for the loan of two fine specimens, collected in Jersey by Miss Dyke-Poore, and from these the above description is taken. They are, unfortunately, both barren, and as this was also the case with the first specimen found, it is impossible to decide with certainty whether the form be entitled to specific rank. In external appearance it is more distinct from all the allied British species than some of these always are from each other. The lax mode of growth, the rounded, irregular outline of the frond, and the very long, and comparatively robust branchlets, are the most striking characters.

Order VIII. LAURENCIACEÆ.

Rose-red or Purple Sea-weeds with rounded, or more or less flattened, branching fronds, not jointed, but sometimes constricted, and composed of many-sided, small cells. Fructification of two kinds on different plants :—1. Pear-shaped spores, in external, egg-shaped, or globular conceptacles ; 2. Tetraspores immersed, without order, among the surface-cells of the branches and branchlets.

Genus XLIII. BONNEMAISONIA.

Frond solid, flattened, much branched, beset with short, slender, awl-shaped, alternate, cilia-like branchlets. Spores pear-shaped, on simple threads, in egg-shaped conceptacles; tetraspores not known.—BONNEMAISONIA, named in honour of M. Bonnemaison, a French naturalist.

The only described species of this genus is pretty widely distributed along the shores of Europe, from the Mediterranean to the Baltic.

Bonnemaisonia asparagoides. Asparagus-like Bonnemaisonia.

Root a small disc. Fronds from a few inches to a foot long, growing either singly or in tufts; stem undivided; branches alternate, becoming gradually shorter towards the top of the frond; branchlets alternate, awl-shaped, about a quarter of an inch long, thickly studded over the whole frond. Spores in ovate conceptacles, which are opposite to, and alternate with, the branchlets.

This beautiful plant has been found in many localities on various parts of our coast. It is very distinct from all other British Sea-weeds, and may be readily recognized, even when seen for the first time. The delicate, cilia-like branchlets, which fringe every part with comb-like regularity, and are always alternate to each other, and opposite either to a conceptacle or to a branch, and the brilliant colour and cellular structure of the frond, are the most obvious characters. Only floating or cast-up specimens of this species are usually to be obtained. It is annual, and grows on rocks at and beyond low-water mark, during summer and autumn.

Genus XLIV. LAURENCIA.

Frond solid, cartilaginous, cylindrical or flattened, pinnately-branched, composed of two strata of cells; those in the centre oblong-angular, arranged longitudinally, those which form the bark, as it were, roundish and minute. Fructification:—1, pear-shaped spores, attached to threads which radiate from a basal placenta, contained, within a cellular pericarp, in ovate conceptacles; 2, tripartite tetraspores arranged irregularly in transverse bands below the tips of the ultimate branchlets; 3, antheridia collected in terminal saucer-shaped receptacles.—LAURENCIA, named in honour of M. de la Laurencie, a French naturalist.

This genus contains several species, which are mostly natives of temperate and tropical latitudes. Dr. Harvey, speaking of the American species, says that they are very difficult to determine, that "indeed in this genus, as in many others, it is often impossible to tell whether we are dealing with species or with sportive forms, without a very careful examination of a number of specimens, or without some knowledge of the circumstances accompanying their development." In a modified degree this remark is applicable to our native species, which, although abundant and well known, are in certain states very difficult to distinguish from each other. In many localities the specimens of this genus are very numerous, and form quite a noticeable feature of the shore vegetation. I remember being much struck with the appearance of the coast of Yorkshire, some miles to the north of Whitby, where a stunted growth of *L. pinnatifida,* turned yellow by exposure, clothed the black rocks between the tide-marks to the extent of many acres, and glistened in the sun as brightly as the yellow lichen on an alpine mountain.

Laurencia pinnatifida. Pepper Dulse.

Frond flattened, cartilaginous, much divided, from one to seven or eight inches high, of a dull purple colour, which is frequently changed to yellow by exposure; branches alternate, toothed, blunt at the tip. Spores in ovate, sessile conceptacles; tetraspores imbedded irregularly in the upper part of the branchlets.

This plant is found in abundance on almost every part of our coast, and is widely distributed in other regions. In certain states it has a hot biting taste, and in consequence of this peculiarity is called *Pepper Dulse,* under which name it was formerly eaten in Scotland. It varies in size, form, and colour, according to the situation in which it grows; those specimens which are found near high-water mark are small and stunted, and usually of a brown or yellow colour, while those which grow in deeper water, and are less exposed, are larger, broader, and dark-brown or purple.

Laurencia cæspitosa. Tufted Laurencia.

Frond cylindrical, narrow, repeatedly pinnate, pyramidal,

from two to six inches high; branches alternate, or rarely opposite; branchlets crowded, tapering at the base, truncated at the tips. Spores not hitherto observed; tetraspores tripartite, numerous, imbedded in the substance of the frond at the tips of the branchlets.

This plant was formerly considered to be only a variety of *L. pinnatifida*, and Dr. Harvey, in his 'Phycologia,' offers it as a species "with some hesitation." It is, without doubt, an instance of the difficulty of determining the species of this genus to which I have already alluded; but while I admit that it is not easy to distinguish between *L. cæspitosa* and certain states of *L. pinnatifida*, and even of *L. obtusa* in dried specimens, I must add that I have always found the general characters of growing or fresh plants sufficiently distinct. *L. cæspitosa* has a cylindrical form and a pyramidal, bushy habit of growth; its fronds are soft and gelatinous, and, in consequence, adhere closely to paper. *L. pinnatifida*, on the other hand, has throughout, a flat spreading habit; its fronds are of a hard, close texture, and cartilaginous and, therefore, much less adhesive when dry. *L. obtusa* has opposite branches, is of a brighter red colour, and differs generally in appearance from this species. *L. cæspitosa* is annual, and grows on stones between the tide-marks. It is common all round our coasts.

Laurencia obtusa. Obtuse Laurencia.

Fronds cylindrical, threadlike, twice or thrice pinnate, from two to nine inches long, growing in large tufts; branches opposite, spreading; branchlets short, spreading, club-shaped, blunt at the tip. Spores in egg-shaped, frequently imperfect, conceptacles, at the tips of the branchlets;

tetraspores tripartite, imbedded in the extreme ends of the branchlets.

This plant is usually parasitic on other algæ; but by no means confined to a single species. It grows within the tide-marks, and is distributed all round our coasts; sparingly in Scotland and the north of England, more frequently towards the south, and in greatest abundance in the Channel Islands. Except in the latter locality it is nowhere so common as *L. pinnatifida*, nor does it assume so many forms as that variable species. It is a summer annual.

Genus XLV. CHAMPIA.

Frond rounded or flattened, branched, tubular, constricted at intervals, divided internally by transverse, membranous walls. Spores egg-shaped or obconical, arranged in clusters on branched threads in conical conceptacles furnished with a terminal pore; tetraspores tripartite, scattered among the surface-cells of the branches and branchlets.

Until lately this genus contained only one species, which was a native of the Cape of Good Hope. Dr. Harvey, in his 'Nereis Boreali-Americana,' has added to it several species which had been considered to belong to other genera: *Chylocladia parvula* of British authors is included among them on the ground that it has ovate or conical conceptacles with a terminal pore, and a sporaceous nucleus which while it resembles that of the typical species of *Champia* differs materially from those of *Chylocladia* and *Lomentaria*.

Champia parvula. The small Champia.

Fronds growing in dense, globose tufts, irregularly

branched, constricted at intervals of once or twice their diameter. Spores in prominent, conical conceptacles scattered over the branches and branchlets. The constrictions are most distinct in the lesser branches and branchlets.

This plant, which was formerly included in the genus *Chylocladia*, grows parasitically on small algæ in tide-pools near low-water mark. It is by no means common, but has been found on most parts of our coasts. It is very abundant in North America.

Genus LXVI. LOMENTARIA.

Frond tubular, constricted at regular intervals, divided internally by transverse, membranous walls. Spores obconical, with very short stalks, arranged on a dense globose nucleus in spherical conceptacles which have no orifice; tetraspores tripartite, scattered among the surface-cells of the branches and branchlets.—LOMENTARIA, from the Greek *lomos*, a cross-line.

This genus, like the last, is comparatively new to the student of British Sea-weeds. It contains the remainder of the species which were formerly called *Chylocladia*. In external appearance it much resembles *Champia*; but the different shape of the conceptacles, the absence of the terminal pore, and the globose arrangement of the spore-nuclei are too important characters to permit the two genera to be combined.

Lomentaria kaliformis. Whorled Lomentaria.

Frond almost gelatinous, hollow, having a pyramidal outline, from six inches to nearly two feet long; main stem undivided, constricted, suddenly tapered at the base, from one-eighth to one-fourth of an inch in diameter; branches arranged in whorls round the constrictions of the stem, spreading, the lower ones longest; branchlets whorled, jointed, beadlike. Spores in globular conceptacles which have a wide pellucid border; tetraspores in the joints of the branchlets.

This species is widely distributed. It grows at various depths, either on rocks, sand, or parasitically on other algæ, and is annual. It varies much in size and form, according to the circumstances of its growth; but its general characters are constant, and may be readily recognised in any but very abnormal specimens. The natural colour is pink; but it is difficult to obtain specimens in that state, except in spring and early summer. Those which are thrown up later in the year are usually much faded, especially the stems and larger branches, and are frequently of a pale yellow or yellow-green colour throughout.

Lomentaria reflexa.   The reflexed Lomentaria.

Frond membranaceous, from two to three inches high; lower branches slender, spreading, irregularly arched, attached at intervals by means of small discs; upper branches undivided, curved, beadlike, tapered towards either end; branchlets few, scattered, spreading or recurved. Spores angular, arranged in a dense mass in spherical conceptacles which have a pellucid border; tetraspores immersed in the upper part of the secondary branches and branchlets.

This is the most rare of all the species of *Lomentaria*, the only recorded British habitats with which I am acquainted being the coast of North Devon near Ilfracombe, and Roundstone Bay, Connemara. It has been found on the coast of Normandy, and, therefore, may probably exist on the adjacent shores of the Channel Islands. It grows on rocks near low-water mark, and is in perfection in summer. Its colour is a dull, dark purple.

Lomentaria ovalis. The oval Lomentaria.

Frond cylindrical, solid, irregularly forked, the lower part naked, the upper beset with simple, elongated, oval, tubular, jointed branchlets. Spores in spherical conceptacles which have a wide transparent border, and are seated on the sides of the branchlets; tetraspores tripartite, scattered among the surface-cells of the branchlets.

This is a very distinct species. It has much the appearance of a land plant, with woody, branched stem and soft, succulent leaves. It grows in the deeper rock-pools near low-water mark, and is in perfection in spring and early summer. It is not very common, but I have found it in several localities on the coast of Wales, Devonshire, the Channel Islands, and elsewhere. It is annual.

Order IX. CORALLINACEÆ.

Frond calcareous, its cells secreting carbonate of lime. Spore-threads in tufts at the base of the conceptacle, separating transversely into four spores at maturity.

Genus XLVII. CORALLINA.

Fronds calcareous, jointed, mostly pinnate; nodes very short, transversely striped. Spore-threads pyriform or club-shaped, four-parted, in ovate or urn-shaped conceptacles, which are formed from the end joint of a branch or

branchlet, and are furnished with a terminal pore.—Co-
RALLINA, from the Latin *coralium*, coral.

This is the most highly organized British genus of this
very peculiar Order of plants. The species which it
contains, and those of the two next genera, were formerly considered to belong to the Animal rather than to
the Vegetable Kingdom, and were accordingly classed
among corals and corallines, and their external appearance fully justified such an arrangement. Their fronds
are composed of ordinary cellular tissue; but this is so
completely coated with carbonate of lime as to give the
whole plant a rigid, stony, coral-like structure, and it
is not until this coating has been removed by soaking
the fronds in acid that the vegetable tissue is visible.

The fronds are divided into articulations, or internodes
of very unequal length and varied shape, these are connected by joints or nodes, which are flexible, being formed
of cellular tissue free from calcareous matter. The
fructification is peculiar, and not yet fully understood.
Organs which are almost perfectly analogous to tetraspores are found in conceptacles which are identical in
form and origin with those which contain the spores of
other genera.

Corallina officinalis. The medicinal Coralline.

Fronds many times pinnate, growing in large numbers
together on rocks or sea-weeds at various depths; articulations of the stem, cylindrical, about twice as long as broad,
of the branches, more or less wedge-shaped, with rounded
shoulders, and of the branchlets linear, cylindrical or
slightly flattened. Fructification in conceptacles formed
either from a terminal articulation, or on the surface of the
frond.

CORALLINACEÆ. 127

This plant is found in large quantities on all parts of our coast. When growing in deep water or in the shade it is of a dull dark purple colour; but when exposed it passes quickly through the various shades of lilac and yellow to milk-white, which is the common colour of cast-up specimens. It is perennial, and may be found at all seasons; but attains its greatest beauty in early spring.

Corallina squamata. The scaled Coralline.

Fronds many times pinnate, growing on rocks in dense tufts of considerable extent; articulations, of the lower part of the stem bead-like, short, with blunt angles, becoming gradually longer, broader, and flatter towards the top, those in the summit of the stem and lesser branches being broadly triangular, flattened, and having very marked acute upper angles. Fructification in conceptacles formed either from a terminal articulation, or on the surface of the frond.

This species is less common than *C. officinalis*, which it much resembles. Externally it is usually taller and more lax than that species, and the articulations of the upper branches are more distinctly triangular. When soaked in acid and examined under a microscope the stripes on the articulations will be found to be more curved and wider apart. It is perennial, and grows on rocks near extreme low-water mark. It is in perfection during summer.

Genus XLVIII. JANIA.

Frond thread-like, jointed, calcareous, forkedly divided; nodes very short. Spore-threads pyriform or club-shaped, four-parted, in urn-shaped conceptacles, which are formed from the axillary articulations of the uppermost branches, are furnished with a terminal pore, and are generally two-horned.—JANIA, from Janira, one of the Nereids.

With the exception of the position of the spore-conceptacles, the characters of this genus are the same as those of *Corallina*, but all our British species may be readily known by their much smaller size and more slender make.

Jania rubens. Red Jania.

Fronds about an inch long, forked, growing in tufts on various sea-weeds; articulations of the branches and branchlets, cylindrical, about four times as long as broad in the middle of the frond, shorter below, longer above. The whole plant of a pale red colour, which is changed to white by exposure. Spores in urn-shaped, horned conceptacles, at the tips of the branchlets.

This is a very common plant, which may always be easily recognized. The stems of *Cladostephus verticillatus* are constantly infested with it, so much so, indeed, that it is sometimes difficult to find a specimen of that plant perfectly free from this parasite, which grows less frequently on several other sea-weeds. It is perennial, and fruits in summer.

Jania corniculata. The horned Jania.

Fronds from one to two inches long, forked, growing in tufts on various sea-weeds; articulations of the branches obconical, flattened, their upper angles sharp and prominent, those of the branchlets cylindrical, thread-like. Spores in urn-shaped conceptacles, in the axils of the upper branchlets.

Dr. Harvey says of this species "that it differs from the more common *J. rubens,* chiefly, if not altogether, in the form of the lower articulations, much as *Corallina squamata* differs from *C. officinalis.* The species has been generally kept up by all authors since the time of Ellis, who first distinguished it. On the British shores it is most common on the southern coasts, while *J. rubens* is found all round the islands."

Genus XLIX. MELOBESIA.

Frond solid, crustaceous, opake, of irregular shape, without joints, having a cellular structure hidden by a calcareous deposit, which may be removed by soaking the plant in acid. Spore-threads erect, oblong, four-parted, in conical conceptacles which have a terminal pore, and are scattered on or sunk in the surface of the frond.—MELOBESIA, named after *Melobosis,* one of Hesiod's sea-nymphs.

In external appearance the species belonging to this genus are extremely unlike anything which we are accustomed to consider a plant. Their mode of growth is somewhat similar to that of Lichens, and certain genera of Fungi; but they differ even from these in the important particular of their calcareous, stony structure. Some of them adhere so closely to the rock on which they grow as to be inseparable from it, and for this reason they are most uncomfortable inmates of an herbarium, and require to be kept in a drawer or box by themselves. Dr. Harvey has figured and described nine species in his 'Phycologia Britannica,' but he appears to consider that four of these are founded on insufficient characters. There is, in fact, a great variety of opinion among writers

K

on the subject, and much observation and study will have to be expended on the genus before the species can be satisfactorily determined. Turning for a moment from the consideration of the details of genera and species to glance at the gigantic processes which are in constant progress around us, it is interesting to note not only how various agencies are employed each for the attainment of its required end, but even how the same object is occasionally accomplished by different means. The vast chalk formation which plays so important a part in the economy of this country was all held in solution in the sea which formerly covered the space now occupied by these islands. It was collected and deposited by myriads of mollusks, coral-polyps, sponges, and sea-weeds,—the latter of kindred nature and structure to those of the Order *Corallinaceæ*. And if we examine the work while it is in progress we shall find that although the agencies are so different, and belong half to the Animal and half to the Vegetable Kingdom, the modes of action are very similar. Thus the animal coral-polyps and the vegetable Melobesiæ, at least certain species, spread themselves over vast tracts of the bottom of the sea, and new coral is formed, or new fronds grow in constant succession, layer above layer: that which is beneath being left to die, decay, and deposit its calcareous remains to form chalk for future generations.

Melobesia calcarea. The chalk Melobesia.

Fronds formed of uniform, globular cells, free, coral-like; branches irregularly divided, spreading, tapering to a blunt point; branchlets simple or forked.

This species is very common, particularly on our

southern and western coasts, and in Ireland; but as it grows in deep water, and is too heavy to be easily thrown up by the waves, good specimens can only be obtained by dredging. In some places the fronds cover a large extent of the sandy bottom of the sea—not attaching themselves, but lying heaped together, strata above strata, only those on the surface being alive.

Melobesia fasciculata. The fasciculate Melobesia.

Fronds formed of uniform, oblong cells, free, coral-like; branches short, cylindrical or flattened, crowded together, their tips truncated, broad, and somewhat concave.

In mode of growth, and some other respects, this species resembles the last, but may be distinguished from it by the form of the tips of the branches. It is of a pale purple colour when fresh, and fades to a dirty white when drying.

Melobesia polymorpha. The many-shaped Melobesia.

Fronds formed of uniform, oblong cells about twice as long as they are broad, thick, stony, closely attached to and encrusting rocks, occasionally rising into short, thick, clumsy branches. Spore-conceptacles minute, extremely numerous.

This species is very common all round our coasts. It grows both in deep and shallow water, and varies in colour accordingly: in the former it is of a dark-purple, while in the latter it is usually of a milky-white.

Melobesia lichenoides. The lichen-like Melobesia.

Fronds formed of uniform, elongate cells about four times as long as they are broad, attached to rocks, free at the edge, variously lobed, spreading, brittle. Spore-conceptacles large, obtusely-conical, scattered or collected in groups.

This plant is well described by its name, as it closely resembles a foliaceous lichen. It varies in colour from dark-purple to yellowish-white.

Melobesia agariciformis. The mushroom-like Melobesia.

Fronds free, globular, thin, much lobed, composed of alternate zones of large and small cells; the former elongated-oblong, about three times as long as they are broad; the latter minute, granular. Spore-conceptacles immersed in the frond, scarcely projecting beyond the surface.

This species grows, without attaching itself, on the bottom of sandy bays of moderate depth. It is of a rosy-pink colour when fresh, and becomes yellowish in drying. It is comparatively rare.

Melobesia membranacea. The membranaceous Melobesia.

Fronds minute, dot-like, very thin, circular, becoming confluent, growing on sea-weeds or *Zostera*. Spore-conceptacles depressed.

Melobesia farinosa. The floury Melobesia.

Fronds minute, irregular in outline, rather thin, pale, growing on sea-weeds. Spore-conceptacles prominent.

Melobesia verrucata. The warty Melobesia.

Fronds thin, expanded, irregularly lobed, pale, growing on sea-weeds. Spore-conceptacles small, very numerous.

Melobesia pustulata. The pimpled Melobesia.

Fronds thick, oblong or lobed, incrusting, smooth, dull purple or green, growing on *Phyllophora rubens*, etc. Spore-conceptacles large, rather prominent, conical, numerous.

The forms referred to under the four last-mentioned names are probably only varieties of some of the preceding species, resulting from circumstances of growth. They are figured and described in the 'Phycologia Britannica,' but only as " reputed species."

Order X. HAPALIDIACEÆ.

Minute calcareous Sea-weeds, composed of a single plane of cellules.

Genus L. HAPALIDIUM.

Frond fan-shaped or lobed, composed of a single series of hyaline cellules.—HAPALIDIUM, from the Greek *apalos*, delicate, and *lithas*, a little stone.

Hapalidium phyllactidium. The fan-shaped Hapalidium.

Parasitical on *Chylocladia clavellosa*.

This very beautiful little Sea-weed was discovered by Professor Allman on a specimen of *Chylocladia clavellosa*, which was growing on the shell of an oyster served up for supper. Dr. Harvey, writing on the subject, takes the opportunity to point out this occurrence as a striking illustration of the necessity for a naturalist to be constantly on the watch.

Order XI. SPHÆROCOCCOIDEÆ.

Purple or Red Sea-weeds with an inarticulate, leaf-like or thread-like frond, composed of many-sided or tubular cells. Fructification of two kinds on different plants:—1. Roundish or elliptical spores formed in bead-like threads, which rise from a basal placenta, in conceptacles with or without a terminal pore. 2. Tetraspores variously disposed.

According to Dr. Harvey, the distinctive character of this Order is the structure of the sporiferous nucleus, which consists of a dense tuft of simple or branched bead-like spore-threads, radiating from a cellular placenta, fixed in the base of the cavity of an external conceptacle.

Genus LI. DELESSERIA.

Frond rose-red, sometimes brown-red, leaf-like, branched, traversed by a distinct midrib, transversely veined. Spores in stalkless conceptacles, either rising from the midrib or from leaflets issuing from it; tetraspores in clusters, called *sori*, on different parts of the frond or leaflets.—DELESSERIA, named after M. Delessert, a distinguished botanist.

The species of this genus are widely dispersed in the northern latitudes of both hemispheres. They are remarkable for their brilliant colours, large size, and elegant form.

Delesseria sinuosa. The sinuous Delesseria.

Frond an oblong or obovate, deeply-cleft, toothed leaf, four to six inches in length, and from one to four inches in breadth, rising from a small disc-root, furnished with a strong midrib, and transverse veins; as the frond grows larger, the first-formed leaf becomes more deeply cleft, till the cutting reaches the midrib; the membrane of the older part decays, and the denuded midribs are again thickened into imperfectly-formed stems and branches, from the sides and top of which spring numerous leaves, similar in form to the original. Spores in hemispherical conceptacles, which are formed from the substance of the midrib, or of one of the transverse veins; tetraspores in sori, which are generally lodged in minute marginal leaflets.

This is one of the handsomest species of this genus, and in form and general appearance resembles a land-plant more nearly than almost any other of our native Seaweeds. In consequence of this resemblance, it is sometimes called the "Oak-leaf Delesseria." The frond varies much in size. In deep-water specimens it is frequently very narrow, and terminates in long, vine-like tendrils. On the coast of North Devon, at Lynton, towards the end of autumn, I found large quantities of very luxuriant specimens cast on shore,—some attached to the stems of *Laminaria digitata*, in company with *Nitophyllum laceratum*, *Rhodymenia palmata*, and *D. alata*, some in large detached masses. I have also found it in the Channel Islands, and other localities. This plant is perennial, and in perfection in summer and autumn.

Delesseria alata. The winged Delesseria.

Fronds four to eight inches high, forked, much branched; branches gradually narrower towards the tips, consisting of a strong midrib or stem, bordered by a flat, wing-like lamina, varying from a line to a quarter, or sometimes nearly half an inch in width; every part of the membrane is furnished with opposite, patent veinlets, connecting the midrib with the margin of the lamina, and themselves connected by pellucid striæ. Spores in spherical conceptacles immersed in the midrib towards the tips of the branches; tetraspores in sori disposed on each side of the midrib, or in proper leaflets, near the tips of the branches.

This is the most common of the *Delesseriæ*, and is found on almost all parts of the British coast. Its colour is generally a bright red, becoming darker in drying. It grows on Algæ, and on the sides of perpendicular rocks facing the sea, near low-water mark. In drying it does not adhere to paper.

Delesseria angustissima. The very slender Delesseria.

Frond four to eight inches high, nearly cylindrical below, compressed above, much and irregularly branched; branches alternate, much divided above, and furnished with forked branchlets. Spores in spherical conceptacles, which are either immersed in the tips of the frond, or in small branchlets, springing from the axils of the upper branches; tetraspores cruciate, in sori on the inflated tips of the branches, or in axillary branchlets.

This plant was first discovered by Mr. Brodie, more than half a century since, and at that time Mr. Turner considered it to be only a variety of *D. alata*. Many years afterwards, Dr. Harvey, "in deference to the repeated protests of Mrs. Griffiths," inserted it in his 'Manual' as a distinct species, "under the temporary name of *Gelidium ? rostratum*, recommending it to the notice of observers, and adding that 'my own opinion on this puzzling matter was not very decided.'" The character in dispute is the membranous wing on either side of the midrib of the frond. Mrs. Griffiths maintained that this was constantly absent in all states of her plant, while it was as constantly present in *D. alata* proper. Professor Agardh appears either to have adopted Mrs. Griffiths's view of the matter, or to have formed a similar opinion, for he speaks of *D. alata* var. *angustissima*, and *D. angustissima*: of the former as a variety, of the latter as a distinct species, and adds, that the two must not be confounded, for their characters are different. 'Dr. Harvey, Dr. Dickie, and others, on the contrary, believe that the absence of the membrane is accidental, and that the supposed *D. angustissima* is but an extremely narrow form of *D. alata*. I have never seen a growing plant of the reputed *D. angustissima*, but I have carefully examined Mrs. Griffiths's and Miss Hutchins's specimens in the British Museum, and in the private collection of Mrs. Gray, and I am bound to confess that these do confirm Mrs. Griffiths's description. There is no membrane, and the appearance of the plant is very different from the normal state of *D. alata*. I am not prepared to decide whether these be sufficient grounds to constitute a title to specific rank; but I feel that they do justify me to maintain a species which already exists in the works of Harvey and Agardh. There can be no doubt that many specimens which are called *D. angustissima* are only narrow forms of *D. alata*; but it does not necessarily follow that Mrs. Griffiths's plant is not a distinct species.

Delesseria hypoglossum. The proliferous Delesseria.

Fronds growing in tufts from a minute disc-root, and

consisting of a primary leaf two to eight inches in length, tapering at each end, with a strong midrib, and indistinct transverse veins; numerous smaller leaves spring from the midrib, and in their turn produce others, until the plant becomes very bushy. Spores very minute, in globose conceptacles formed in the substance of the midrib; tetraspores in linear sori, on each side of the midrib of the small leaves.

This plant is less common on our shores than either *D. sinuosa* or *D. alata*, and English specimens are generally small and stunted. On the west coast of Ireland, especially at Bantry Bay, it grows in great abundance, and is of a very large size, frequently attaining a height of eight or nine inches. It is also of a brighter colour, and more delicate texture, but even here small specimens occasionally occur among their more luxuriant brethren. This is a summer species, and annual.

Delesseria ruscifolia. The obtuse-leaved Delesseria.

Fronds from two to four inches high, growing in tufts; primary leaves linear-oblong, undivided, blunt at the tip, often somewhat waved and curled, having a strong midrib, from which springs a secondary series of smaller leaves, which again from their midribs produce a third series in the same manner, and so on; all the leaves are similar; the membrane of which they are formed is composed of minute, densely-packed, angular cellules, and is traversed by anastomosing, branched, jointed veinlets, which run obliquely from the midrib to the margin of the leaf. Spores in conceptacles seated on the midrib, generally near the tips of the younger leaves; tetraspores in linear sori, disposed on each side of the midrib.

Some of the forms of this plant resemble *D. Hypoglossum* so closely, that the difference is scarcely perceptible to the naked eye; but when the two plants are examined under the microscope, the distinctive characters which separate them are very easily perceived. In *D. ruscifolia* the membrane is composed of very minute, closely-packed cellules, and the veinlets are very distinct, while in *D. Hypoglossum* the cellules are comparatively large and wide apart, and either there are no veinlets, or they are very faint. This species is annual, and to be found from spring to the end of autumn.

Genus LII. **NITOPHYLLUM.**

Frond rose-red, membranaceous, irregularly cleft, without a midrib, either veinless or with a few slender, branched, vanishing veins near the base. Spores elliptical or roundish, in globose, stalkless conceptacles scattered over the surface of the frond; tetraspores grouped in spots, or sori, also scattered over the frond.—NITOPHYLLUM, from the Latin *niteo*, to shine, and the Greek *phyllon*, a leaf.

This genus has many characters in common with *Delesseria*, the principal difference between them being the form of the frond, which is irregular and unsymmetrical in *Nitophyllum*, and regular and leaf-like in *Delesseria*. All the species of the latter genus, moreover, have distinct midribs, while in those of the former, that organ is only represented by very slender, often vanishing veins, which are altogether absent in some species. All the species are annual, and flourish during summer and autumn. As a rule they grow in deep water, and can only be observed *in situ* during extreme low-tides. They are, however, constantly thrown on shore, either on the stems of *Laminariæ* or detached.

Nitophyllum punctatum. The dotted Nitophyllum.

Fronds growing in tufts from a small disc-root, from four to twenty inches long, and nearly as broad, oblong, delicately thin, destitute of veins, either regularly divided or cleft into two or three principal segments, fringed at the edge with dichotomous lobules. Spores contained in large hemispherical conceptacles, which are thickly scattered over the whole frond; tetraspores tripartite, grouped in large, dark red, oblong spots.

The above description applies strictly to the normal, or typical, state of the species. In the 'Phycologia Britannica,' Dr. Harvey has described three varieties, which, though differing from each other very considerably in form and general appearance, are not sufficiently distinct to be separated into species. The first variety β. *ocellatum* is divided to the base into linear, dichotomous lobes, with a perfectly even flat margin. In γ. *crispatum* the frond is thicker and of a darker colour, from half an inch to an inch broad, and from six to eight inches long, dichotomously divided, and strongly curled at the margin. Variety δ. *Pollexfenii* is also thicker than the typical form, wedge-shaped at base, variously lobed, and has a rounded margin. Variety ε. *fimbriatum* is very thin, without fruit, roundish, the margin cut into minute forked lobes about a line in breadth. There are other less distinct varieties intermediate between all of these, which gradually connect the extreme broad and narrow forms in one unbroken series, and clearly prove that all belong to the same species, altered only by varying circumstances of growth. *Nitophyllum punctatum* is found on many parts of the English coast, but nowhere in great abundance, and very seldom of a large size. The finest specimens that I have seen from English habitats are some that were collected at Plymouth by Mr. Gatcombe. On the west coast of Ireland, particularly in Roundstone Bay, it grows in large quantities and in great luxuriance. In this state, its broad, delicate, pink fronds, and the elegant arrangement of its dotted fructification, entitle it to be considered one of the most beautiful of our native Sea-weeds.

Nitophyllum Hilliæ. Miss Hill's Nitophyllum.

Frond fan-shaped, thickish, very irregularly divided, sometimes nearly simple, sometimes cleft into a few broad segments, and sometimes deeply cut into ribbon-like laciniæ, proliferous from the margin and much waved, veined at the base. Spores in large hemispherical conceptacles, which are irregularly scattered over the frond; tetraspores in minute, dot-like spots, also scattered over the frond.

This species is less common than the preceding. It grows on the shady sides of deep, tidal pools, near low-water mark, and is in perfection in summer and autumn. Its most obvious characteristics are its thicker substance, and large scattered spore-conceptacles. I am also indebted to Mr. Gatcombe for fine Plymouth specimens of this species.

Nitophyllum Bonnemaisonii. Bonnemaison's Nitophyllum.

Frond hand-shaped, dichotomously divided, two to four inches in length, expanding from a short cylindrical stalk, which springs from a disc-root; veins sometimes confined to the base, sometimes extending considerably up the segments. Spores in small, not very prominent conceptacles scattered over the frond; tetraspores oblong or roundish, abundantly scattered over the frond.

Like the other species of this genus, *Nitophyllum Bonnemaisonii* varies very much in form. Some fronds are cleft nearly to the base, others are scarcely divided at all; some are entire, and some proliferous at the margin. It grows in deep water, and is sometimes cast on shore on the stems of *Laminaria digitata*.

Nitophyllum Gmelini.   Gmelin's Nitophyllum.

Frond from two to six inches high, expanding from a short stalk, which springs from a small disc-root, more or less deeply cleft into waved lobes; veins at the base of the frond distinct, gradually becoming fainter upwards. Spores in hemispherical, depressed conceptacles, which are generally formed near the margin of the frond, or in the upper lobes; tetraspores always placed just within the margin of the frond.

This species grows on rocks and large sea-weeds, at and beyond low-water mark. It is in perfection in summer, and occurs most frequently on the south coast of England, the west of Ireland, and the Channel Islands. It may be known by its marginal *tetraspores*, and peculiar crisp texture.

Nitophyllum laceratum.   The jagged Nitophyllum.

Fronds four to six or eight inches long, dichotomously divided, sessile, or having a very short stalk with a disc-root, which sometimes throws out creeping fibres; the lobes are of a thin delicate substance, linear or wedge-shaped, with a curled or dentated margin; the basal veins are distinctly marked, and in some specimens extend through all the branching of the frond. Spores forming a chain round a basal placenta, in depressed spheroidal conceptacles, which are either disposed on the margin of the frond, or on leafy processes issuing from it; tetraspores grouped in minute, oblong spots on the margin of the frond, or on marginal processes.

This is the most variable, as well as the most common of all the British species of *Nitophyllum*. In some specimens the divisions of the frond are almost thread-like, while in others they are so broad as to approach the form of *Nitophyllum Gmelini*. These differences arise, in a great measure, from the peculiar circumstances in which the plants grow,—whether on rocks or the stems of *Laminariæ* in deep water, or, as I have sometimes found them, in shallow, tidal pools. The only species with which this can be easily confounded is *N. Gmelini*, and from that it may be distinguished by its darker purple colour, softer texture, more distinct, longer veins and proliferous margins. The texture of the frond is so delicate when dry that it is apt to crack and part from the paper on the least exposure to the air. Great care should therefore be taken of specimens in the herbarium. A curtain of tissue paper gummed neatly above the specimen, so as to fall over it, is the best protection that I can suggest.

Nitophyllum uncinatum. Clawed Nitophyllum.

Frond membranaceous, without stalk or veins, divided into linear lobes, whose tips are incurved. Spores in depressed, spheroidal conceptacles, which are usually to be found near the margin of the frond; tetraspores in nearly solitary sori, under the tips of the lobes.

I insert this species in deference to the authority of Dr. Gray's Handbook, from which I have copied the description almost verbatim. Dr. Gray speaks of it as identical with the variety of *N. laceratum* figured in the 'Phycologia Britannica,' plate 267; but he also says that it has no veins, which the figure referred to certainly has. I have examined several specimens in my own collection, but have not yet convinced myself that they are sufficiently distinct to be considered more than a very peculiar form of *N. laceratum.*

Nitophyllum versicolor. The changeable Nitophyllum.

Fronds broadly fan-shaped, dichotomously divided, veinless, expanding suddenly from a tuberous cylindrical stem, about one and a half to two inches high, and one to two lines thick; the tips of the frond, and sometimes the side margin, are much thickened, and produce oblong, fleshy excrescences, which, as they advance in age, lengthen into irregularly branched, cellular, cylindrical filaments. Fructification unknown.

The slightest contact with fresh water changes this plant from red to bright orange, but the original colour is restored and retained when the plant is dry. It is very rare, and the only specimens hitherto found have been thrown up from deep water on the coasts of Ireland and Devonshire.

Genus LIII. **CALLIBLEPHARIS.**

Frond flat, membranaceous, irregularly cleft, fringed with marginal lobes, without veins or midrib. Fructification of two kinds on different plants :—1, necklace-like threads of spores, springing in dense tufts from an elevated basal placenta in stalkless conceptacles on the marginal processes of the frond; 2, zonate or tripartite tetraspores embedded among the surface cells.—CALLIBLEPHARIS, from the Greek *kalos*, beautiful, and *blepharis*, an eye-lash.

Calliblepharis ciliata. The ciliate Calliblepharis.

Frond six to twelve inches long, rising from a short, cylindrical stem, with a branching, fibrous root; substance crisp, rigid, the surface of the frond sometimes smooth, sometimes covered with cilia. In some specimens the frond consists of a simple lance-shaped leaf, with awl-shaped processes on its margin; in others the main frond is variously cleft, and furnished at the edges with lance-shaped segments, from whose margins issue similar processes, which are sometimes prolonged into lobes. Spores in conceptacles, lodged in the marginal processes; tetraspores collected in cloud-like patches in various parts of the frond.

This plant, better known to the marine botanist by its old name *Rhodymenia ciliata*, has been transferred to its present position, on account of the structure of its spore nucleus and the zonate division of the tetraspores. It is found on all our coasts from north to south, and fruits in winter.

Calliblepharis jubata. The cirrhose Calliblepharis.

Fronds tufted, rising with a cylindrical stem from a root composed of densely-matted, branching fibres; branches irregularly pinnate, of very variable form, clothed throughout with filiform cilia, which in some specimens are produced into tendrils three to six inches long, and generally hooked at the point; these cilia frequently twine round neighbouring sea-weeds, and connect them into an inextricable tangle. Spores in spherical, stalkless conceptacles on the sides of the cilia; tetraspores oblong, zonate, embedded in the cilia.

This species is closely allied to the preceding, but differs from it in being of a duller colour and more flaccid substance, and in having its tetraspores confined to the cilia. It fruits and is in perfection in summer.

Genus LIV. GRACILARIA.

Frond thread-like or flat, of a horn-like, fleshy substance, forked; the inner cells very large, empty, or full of granular matter; those of the surface minute, forming densely-packed vertical threads. Fructification:—1, spores formed in the upper joints of densely-tufted, forked, necklace-like threads, which radiate from a raised basal placenta, in hemispherical or conical, stalkless conceptacles furnished with a terminal pore; 2, oblong, cruciate tetraspores scattered among the surface cells of the branches and branchlets.—GRACILARIA, from the Latin *gracilis*, slender.

The species of this genus are widely dispersed, both in warm and cold latitudes. They produce, when boiled, a tasteless gelatine, which, when properly seasoned, is palatable, and considered wholesome.

Gracilaria multipartita. The much-divided Gracilaria.

Fronds four to twelve inches long, flat, varying from half a line to half an inch in width, deeply cleft in an irregularly-forked or palmate manner, with a thin, spreading disc-root. Spores in large, conical conceptacles, which are depressed at the top, and spread abundantly over the frond; tetraspores tripartite, scattered over the whole frond.

This species is annual, and grows, during autumn, on rocks or mud at and beyond low-water mark. When freshly gathered, the fronds are tender and very brittle, but they become tough in drying, and adhere closely to paper. They are of a dull purple colour, changing to green on exposure to the air and fresh water. Specimens have been found in several places on the coast of Devonshire, but this must still rank as one of the most rare of our native Sea-weeds.

Gracilaria compressa. The compressed Gracilaria.

Frond from six to twelve inches long, of a pale pink colour, very brittle, alternately or almost forkedly branched; branches long, mostly simple, tapering to a fine point. Spores minute, at the tips of threads, radiating from a central point in prominent, egg-shaped, stalkless conceptacles; tetraspores roundish, tripartite, irregularly dispersed among the surface cells.

This species is also rare, and there are but few recorded habitats. It grows on corallines, in deep water, during summer, and is annual. My specimens were found amongst rejectamenta at Swanage, a neighbourhood where many rare Sea-weeds are to be obtained.

Gracilaria confervoides. The conferva-like Gracilaria.

Fronds cylindrical, cartilaginous, rising one or several from the same base, varying from three to twenty inches in length; branches long, thread-like, nearly simple, attenuate;

branchlets few, tapering at each end. Spores minute, in large, roundish, stalkless conceptacles plentifully scattered over the branches; tetraspores minute, embedded in the surface cells of the branches.

This species is perennial, and attains its greatest luxuriance during summer and autumn. It is found plentifully all round our coasts, and is readily distinguished by its prominent spore-conceptacles, which have the appearance of knots on the slender branches. The colour is a dull purple, which fades to white when the plant grows in shallow places, or is exposed to the air and light. When dry it is rigid, and does not adhere closely to paper. Dr. Harvey speaks of a variety found in American waters, which is destitute of branches, and reaches the length of six feet. I am not aware that anything approaching this has been found on our coasts.

Genus LV. SPHÆROCOCCUS.

Frond horny, compressed, linear, two-edged, forkedly branched, composed of three series of cells; those in the centre, which form a kind of internal rib, fibrous and densely packed, the next series large and many-sided, and those on the surface minute and thread-like. Spores minute, arranged on a central placenta in spherical conceptacles; tetraspores, according to Kützing, zonate, scattered through the substance of the frond.—SPHÆROCOCCUS, from the Greek *sphaira*, a sphere, and *coccos*, fruit.

Sphærococcus coronopifolius. The buck's-horn Sphærococcus.

Fronds from a few inches to a foot or more long, much branched; main stems thickened and two-edged below, becoming thinner and flatter in their upper parts, irregularly divided; the upper branches once or twice forked, ending in fan-shaped, many-cleft branchlets; the branches and branchlets fringed with short, slender, pointed cilia. Spores in spherical conceptacles, embedded in the cilia; tetraspores not known.

The bright red colour and coral-like form of this handsome species are very obvious characters, and there is no other British Sea-weed for which it can be mistaken. It grows on rocky shores, at or beyond low-water mark, during summer and autumn, and is perennial. Unless a dredge be used, only washed-up or floating specimens can be obtained, but these are always sufficiently numerous. This plant is moderately abundant on our south and west coasts, but rare in Scotland. It does not adhere very closely to paper, and shrinks much in drying.

ORDER XII. GELIDIACEÆ.

Purple or Red Sea-weeds with a cartilaginous or horny, opake frond, without joints, composed of elongated, hair-like filaments. Spores attached to slender threads or to a fibrous placenta, in conceptacles, which are irregularly immersed in the frond.

Genus LVI. **GELIDIUM.**

Frond gristly, linear, flattened, irregularly branched, composed of three strata of cells: those in the centre very long, and densely interwoven; the next series small, many-sided, set in diverging lines; those on the surface minute, arranged in necklace-like threads at right-angles with the axis. Spores pear-shaped, in two-celled conceptacles immersed in the branchlets; tetraspores cruciate, embedded, in irregular clusters, near the ends of the branchlets.—GELIDIUM, from the Latin *gelu*, frost.

The species which compose this genus are remarkable for the variety of forms which they assume, and for their wide geographical range. They are found on all the tropical and temperate shores of the Atlantic and Pacific Oceans. Two species have been hitherto included in works on British Sea-weeds, but the claims of *G. cartilagineum* are so slight that I have felt compelled to omit it. It is a common weed at the Cape of Good Hope, whence it is frequently brought by sailors and thrown overboard in British waters, and the specimens picked up on our coasts have, no doubt, been derived from this source.

Gelidium corneum. The horny Gelidium.

Frond flattened, rigid, several times pinnate; pinnæ narrowed at the base, linear, entire, obtuse. Spores in conceptacles below the tops of the branches; tetraspores in club-shaped very obtuse pinnæ.

The varieties of this species are more numerous than those of any other of our native Sea-weeds. Dr. Harvey has described no less than thirteen, and it would be very easy to increase the number, were it desirable to do so. The differences consist chiefly in the size, breadth, and branching of the frond, but there can be no doubt of the specific identity of the whole series. I shall therefore content myself with enumerating the varieties mentioned by Dr. Harvey, and recommending my readers to make as complete a collection as possible of all the forms which they may find. The varieties are :—

G. CORNEUM, var. β. *sesquipedale*, Grev.
,, var. γ. *pinnatum*, Grev.
,, var. δ. *uniforme*, Turn.
,, var. ε. *capillaceum*, Turn.
,, var. ζ. *latifolium*, Grev.
,, var. η. *confertum*, Grev.
,, var. θ. *flexuosum*, Harv.
,, var. ι. *aculeatum*, Grev.
,, var. κ. *abnorme*, Grev.
,, var. λ. *pulchellum*, Turn.
,, var. μ. *claviferum*, Grev.
,, var. ν. *clavatum*, Grev.
,, var. ο. *crinale*, Grev.

ORDER XIII. SPONGIOCARPEÆ.

Brown-red Sea-weeds with cartilaginous, cylindrical, branching fronds composed of interlaced filaments closely set in firm gelatine. Spores large, obconical, radiating from a central point, in external, wart-like conceptacles; tetraspores oblong, cruciate, formed in the outer filaments of the upper branches.

Genus LVII. **POLYIDES.**

Root an expanded, fleshy disc. Frond cylindrical, cartilaginous, forkedly branched, composed of three series of fibrous cells, springing one from the other: the cells of the first, or central series, which occupy about half the diameter of the frond, are cylindrical, branched, longitudinal, and densely packed; those next to them, which form the intermediate series, are large, oblong, coloured, and set in forked fibres, which curve outwards, and pass gradually from an almost erect to a horizontal position; and those which compose the external, or bark series, are minute, oblong, forked, and perfectly horizontal. Spores in wart-like excrescences; tetraspores cruciate, formed in the upper branches.—POLYIDES, from the Greek *polus*, many, and *idea*, form.

Polyides rotundus. The round Polyides.

Fronds growing several from the same base, from two to six inches high, forkedly divided, of a cartilaginous substance, and dark red-brown colour, not adhering to paper.

This species is the only one contained in the Order, and is very distinct from all the British Sea-weeds except *Furcellaria fastigiata,* which it resembles so closely in external appearance that it is difficult to believe that the two are really so different in structure. There is, fortunately, one very simple character by which they may be identified: the root of *P. rotundus* is a small disc; that of *F. fastigiata* is fibrous. The former is, moreover, much less common than the latter. *P. rotundus* grows on rocks in tide-pools and in deeper water. It is perennial, and attains perfection in winter; but may be found at other seasons.

Order XIV. SQUAMARIÆ.

Lichen-like, incrusting or horizontally expanded, Red-brown Sea-weeds, rooting by the under surface, composed of vertical filaments closely set in firm gelatine. Spores in necklace-like strings, lodged in wart-like excrescences, formed of vertical filaments.

Genus LVIII. PEYSSONELIA.

Frond membranous or leathery, horizontally expanded, and attached by fibrils emitted from the lower surface, composed of two strata of cellules: those of the lower stratum horizontal, elongated, cylindrical, arranged in radiating fibres, which are placed close together, and form a membrane; those

SQUAMARIÆ. 153

of the upper vertically elongated, also arranged in fibres at right angles to those of the lower stratum. Fructification lodged in depressed warts, which are irregularly scattered over the surface of the frond:—1, roundish spores, in necklace-like strings; 2, oblong, cruciate tetraspores.— PEYSSONELIA, in honour of J. A. Peyssonel.

Peyssonelia Dubyi. Duby's Peyssonelia.

Frond membranaceous, from half an inch to an inch in diameter; when young, round; in age becoming irregularly-shaped, with a waved margin; the upper surface finely striate, the lines radiating from the centre to the edge of the frond. Fructification in spongy warts, scattered over the frond.

This is a species which may easily escape the eye of a collector. It forms a thin, reddish skin on the surface of rocks or shells, and can only be separated from them with difficulty. Sometimes a portion only of the frond is attached, the ends being left free.

Genus LIX. HILDENBRANDTIA.

Frond spreading, in a leathery skin, over rocks and stones, its substance formed of closely-packed, vertical fibres. Spores in warts on the surface of the frond; tetraspores oblong, zonate, minute, sunk in the frond.

Hildenbrandtia rubra. The red Hildenbrandtia.

Frond at first circular, but becoming irregularly lobed with age, varying in colour from blood-red to dark brown.

This is an obscure plant, which, although of frequent occurrence, is but little known. To the casual observer it appears to be but a dark stain on the rock on which it grows, and to which it adheres so closely that it is impossible to detach a perfect specimen. In structure and fructification there is a close affinity between this genus and *Melobesia;* but the coating of carbonate of lime, which distinguishes the latter genus, is altogether absent from *Hildenbrandtia.*

Genus LX. PETROCELIS.

Frond spreading, in a leathery skin, over rocks, composed of simple, jointed fibres set in gelatine. Spores unknown; tetraspores cruciate, formed of the middle joint of the fibres, which compose the frond.—PETROCELIS, from the Greek *petros,* a rock, and *chele,* a claw.

This genus was formerly considered to be identical with *Cruoria.* The separation is founded on the different arrangement of the tetraspores.

Petrocelis cruenta. The blood-red Petrocelis.

Fronds spreading, in smooth, circular or irregular patches, on rocks, gelatinous, composed of vertical filaments.

This is another obscure species, destitute of external beauty, and not likely to prove interesting to young collectors.

Genus LXI. CRUORIA.

Frond expanded, composed of vertical, jointed fibres, which are very closely packed, and surrounded by fluid gelatine. Spores unknown; tetraspores zonate, borne on transformed side-branches of the fibres.—CRUORIA, from the Latin *cruor,* blood.

This genus, as now constituted, is comparatively new to the British flora, but since it has been described numerous specimens of both the species that it contains have been collected in various localities, particularly in Scotland. I am, therefore, inclined to believe that unobtrusive habit, rather than rarity, has caused it to be hitherto overlooked.

Cruoria pellita. The skin-like Cruoria.

Fronds spreading over rocks in indefinite red patches; fibres thick at the base, tapering upwards, repeatedly forked. Tetraspores on the lower branches of the fibres.

This plant grows between the tide-marks, and appears to prefer a cold climate.

Cruoria adhærens. The adhering Cruoria.

Fronds spreading over rocks, in indefinite, purplish or olive patches; fibres tapering to either end, simple, or once or twice forked. Spores unknown; tetraspores zonate, on the lower forkings of the fibres.

This species differs from the last in the colour of the frond, and in the branching and form of the fibres; but Professor Agardh appears to doubt whether these characters are sufficient grounds for the formation of two species.

Genus LXII. ACTINOCOCCUS.

Fronds globose, very minute, of a red colour, parasitic, composed of necklace-like fibres set in gelatine. Tetraspores cruciate, formed from the joints of the fibres, several on the same fibre.—ACTINOCOCCUS, from the Greek *aktis*, a ray, and *coccos*, a fruit.

Actinococcus Hennedyi. Hennedy's Actinococcus.

Fronds of a deep red colour, about as large as a poppyseed, growing on *Laminaria digitata*, composed of closely-packed necklace-like fibres set in gelatine. Tetraspores cruciate, bright scarlet.

This tiny plant was first discovered by Mr. R. Hennedy, at Cumbrae, in 1852, and was described by Dr. Harvey in the 'Natural History Review' for 1857.

Order XV. HELMINTHOCLADIÆ.

Rosy or Purple, cylindrical, gelatinous or gelatino-membranaceous Sea-weeds composed of filaments set in loose gelatine. Spores minute, roundish, borne on branching threads, which radiate in a spherical form from a central point and are immersed in the frond without conceptacles; tetraspores, when present, formed in the terminal cellules of the outer filaments.

Genus LXIII. HELMINTHORA.

Frond cylindrical, gelatinous, elastic, much branched, composed of two series of filaments set in loose jelly; those in the centre, which form the axis, are parallel, longitudinal, jointed, and decrease in size outwards; those forming the outer part of the frond, or periphery, are bead-like and forked, and issue horizontally from the axis. Spore-threads numerous, club-shaped, embedded in the outer filaments of the fronds; tetraspores unknown.—HELMINTHORA, from the Greek *elmins*, a worm, and *thoros*, a seed.

This genus is new to collectors of British Sea-weeds. It has been established by Professor Agardh, and adopted by Dr. Harvey for the reception of what was formerly *Dudresnaia divaricata*.

Helminthora divaricata. The divaricate Helminthora.

Fronds densely tufted, thread-like, pale red; stem simple or forked; branches opposite or alternate, horizontal, much divided; branchlets numerous, divaricate, scattered, obtuse. Spores on club-shaped threads, embedded in the frond; tetraspores unknown.

This plant is annual, and grows on stones and seaweeds during summer and autumn, at and beyond low-water mark. It is widely distributed in the northern latitudes of America and Europe, and is found in many places on the coasts of Great Britain.

Genus LXIV. NEMALION.

Frond cylindrical, gelatinous, elastic, forked, composed of filaments set in gelatine; those of the axis simple, longitudinal, closely twisted into a cord-like column, surrounded by other oblique, communicating filaments, from which issue the horizontal, necklace-like, fastigiate, forked filaments of the outer part of the frond, or periphery. Spore-threads club-shaped, embedded in the outer filaments of the frond; tetraspores, according to Agardh, "formed in the terminal cells of the peripheric filaments, triangularly divided, with prominent sporules."—NEMALION, from the Greek *nema*, a thread, and *leion*, a crop.

Nemalion multifidum. The many-cut Nemalion.

Fronds worm-like, six to ten inches long, as thick as a crow's quill, or thicker, of a dull purple colour, forked near the base, and repeatedly at long intervals upwards; root an expanded fleshy disc. Spore-threads embedded in the frond; tetraspores formed in the terminal cellules of the outer filaments.

This species grows on rocks and shells near low-tide mark, and occasionally, when the rocks are formed of granite, in shallow shore-pools. It appears to prefer situations where it is exposed to the air when the tide goes out and where it is washed and beaten by the waves. It is common on the western coasts of Scotland and Ireland, and occurs less frequently in several English localities, chiefly on the south coast. This is the only plant of the Order *Helminthocladiæ* on which tetraspores have been found.

Genus LXV. HELMINTHOCLADIA.

Frond cylindrical, elastic, composed of filaments, set in gelatine; those of the axis simple, longitudinal, loosely interwoven into a cord-like column, surrounded by other oblique, communicating filaments, from which issue the horizontal necklace-like, forked filaments, which compose the periphery. Fructification, masses of spores seated among the filaments of the periphery; tetraspores unknown.— HELMINTHOCLADIA, from the Greek *elmins*, a worm, and *klados*, a branch.

Helminthocladia purpurea. The purple Helminthocladia.

Fronds from eight inches to two feet or more in length, tapering from the centre towards either end; stem mostly undivided; branches opposite or alternate, irregular, numerous; branchlets slender, undivided; colour varying from a deep purple-red to a dull pink, rapidly given out in fresh water, and becoming brown in drying. Root a small disc. Fructification, globular masses of spores among the threads of the periphery.

In the ' Phycologia Britannica ' this plant is figured

and described as "*Nemaleon purpureum*," but in more recent works a new genus has been formed for its reception, where at present it stands alone. It is a summer annual, and is found among *Zostera* on sandy shores. It is much more rare than *Nemalion multifidum*, but occasionally considerable quantities are thrown on shore. It would almost appear that its growth is dependent on temperature or weather, and that it is, therefore, comparatively abundant in some seasons and very scarce in others.

Genus LXVI. SCINAIA.

Frond rounded, forked, gelatinously membranaceous, filled with mucilage, traversed by a fibrous axis, from which slender, forked filaments radiate horizontally, and unite at their tips to form the external membrane of the frond. Spores pear-shaped, borne on fastigiate, jointed threads arranged in globose clusters just within the walls of the frond; tetraspores unknown.—SCINAIA, probably from the Greek *Schinis*, a name of *Aphrodite*.

Scinaia furcellata. The forked Scinaia.

Frond cylindrical, tender, two to four inches long, varying much in diameter, many times forked, level-topped, with a rounded outline when laid out. Spores pear-shaped; tetraspores unknown.

This rare and interesting plant, which was formerly called *Ginannia furcellata*, varies much in size and substance. The wide specimens are more delicate in texture, and of a paler colour, than the narrow. The branches are bag-like, and the gelatine with which they are filled is exuded in drying, and causes the plant to adhere very firmly to paper. The cord-like axis is very distinct in some specimens, and has the appearance of a midrib. The spores are described by Professor Agardh as naked; but Dr. Harvey has since raised a doubt on this point by discovering, with the aid of a high magnifying power, what he believed to be the remains of a very delicate, membranous pericarp.

Order XVI. WRANGELIACEÆ.

Rose-red, thread-like Sea-weeds, with or without joints, traversed by a single-tubed jointed axis. Fructification:— 1. *Spores formed in the terminal cell of branching threads, which radiate in naked clusters, either from a fixed point or round minute side-branchlets.* 2. *Tetraspores formed of branchlets shortened to a single cell, naked, not present in all the genera.*

Genus LXVII. WRANGELIA.

Frond thread-like, much branched, jointed, one-tubed; internodes of the axis naked, or covered with minute cellules; nodes clothed with opposite or whorled, jointed branchlets. Spore-clusters terminal in a nest of fibrous branchlets; tetraspores naked, spherical, triangularly divided, seated on the sides of the branchlets.—WRANGELIA, named in honour of Baron Von Wrangel, a Swedish naturalist.

The species of this genus are not numerous, and are mostly natives of warm or temperate latitudes. They were formerly placed in the Order *Ceramiaceæ*, between the *Griffithsiæ* and *Callithamnia*, which they externally resemble. They are removed to their present position on account of important differences in their mode of fructification.

Wrangelia multifida. Many-cut Wrangelia.

Fronds jointed throughout, growing in tufts, from four to eight inches high; stem generally undivided; internodes formed of a single, cylindrical, thick-walled cell, filled with carmine endochrome; branches opposite, inserted just below each node; branchlets opposite or whorled, incurved, multifid, pervading every part of the frond. Spore-clusters arranged, several together, in globular masses of involucral branchlets at the tips of short branches; tetraspores roundish, tripartite, on the lower part of the branchlets.

This very delicate and pretty species grows in rock-pools near low-water mark. It occurs in several localities on our south coast, in the west of Ireland, and in the Channel Islands. I have frequently gathered it in Jersey, but never in large quantities. Mr. John Gatcombe, of Plymouth, from whom I recently received some very interesting specimens in a young state, mentions a fact connected with this plant that has not, so far as I am aware, been observed before. He writes, "*Wrangelia* can be detected immediately when first gathered by its peculiar smell."

Genus LXVIII. **NACCARIA.**

Frond much branched, composed of many-sided cells of different sizes, those in the centre large, those on the surface minute; tubular axis slender, " girt with rounded angular cells, formed with the decurrent fibres ;" branches forked. Spore-threads whorled round the branchlets, which become spindle-shaped as the spores mature ; tetraspores unknown.—NACCARIA, named in honour of F. L. Naccari, an Italian botanist.

This is another genus whose affinities have been

rudely dealt with by modern science. It was formerly classed among the *Gloiocladeæ* on account of its gelatinous frond; but this character is now considered to be of less importance than the arrangement of the spores, and hence the altered position of the genus.

Naccaria Wiggii.   Wigg's Naccaria.

Frond cylindrical, from a few inches to nearly a foot high; branches alternate or irregular; branchlets short, spindle-shaped. Spores oblong, naked, on threads which are whorled round the branchlets; tetraspores unknown.

This plant is annual, and grows in deep water. It is of a beautiful bright red colour, and somewhat gelatinous. The structure of the lesser divisions of the frond is very interesting, and should be examined under the microscope, which will reveal the whorls of delicate, jointed fibres that compose the branchlets, and amid which the spores nestle. It is, moreover, on these minute details that the distinctive characters of the genus chiefly depend. Specimens of this species are cast up in many localities on our coast, but nowhere in great abundance. I have found more in Jersey than anywhere else, and they were usually floating and in good condition.

Genus LXIX.   ATRACTOPHORA.

Frond nearly oppositely branched, composed of large and small cells; tubular axis broad, "closely barked with decurrent articulated fibres;" branches pinnate. Spore-threads whorled round the branchlets; tetraspores unknown.— ATRACTOPHORA, from the Greek *atractos*, a spindle, and *phero*, to carry.

This genus and *Naccaria*, so far at least as the British species were concerned, were formerly combined. Professor Agardh, in his 'Species Algarum,' has separated them, and as I have been following the arrangement of that work as nearly as possible, I have done so in this particular. The words between inverted commas in the descriptions of these two genera are from Dr. Gray's 'Handbook of British Water-weeds.'

Atractophora hypnoides. Hypnum-like Atractophora.

Fronds slender, much branched, from two to four inches high, and about the same breadth; branches alternate, long, spreading; branchlets slender, jointed, whorled with minute, necklace-like, forked fibres. Spores naked, set among the whorled fibres of the branchlets; tetraspores unknown.

This is a recent addition to our list of British seaweeds. It was first discovered in Jersey by Miss Turner, and has been subsequently found at Exmouth by Mrs. Gulson. It is a very beautiful and distinct plant, and will prove a rich prize to those collectors who may be so fortunate as to find it either in the above or in new localities.

ORDER XVII. RHODYMENIACEÆ.

Purple or Red Sea-weeds with a flat, compressed, or thread-like, membranaceous frond, without joints, composed of many-sided cells, the surface cells forming a continuous coating. Fructification: — 1. *Spores in necklace-like, branching threads issuing from a placenta, massed together without order at maturity, lodged in external conceptacles.*

M 2

2. *Roundish or oblong tetraspores, variously parted, either dispersed among the surface cells, or collected in definite clusters, or in proper leaflets.*

"This Order," writes Dr. Harvey, "has recently been proposed by Professor J. G. Agardh, to include a few genera which, on account of the very different structure of their conceptacular fruit, he has rejected from the *Sphærococcoideæ*,—a measure rendered necessary by the new principles of arrangement developed by that author. These plants, however, so closely resemble the genuine *Sphærococcoideæ* in external habit, and even in the internal structure of the stem and leaves, that recourse must sometimes be had to an accurate microscopic analysis of the contents of the conceptacle, before the student can ascertain the proper place in the system of the plant under examination."

Genus LXX. MAUGERIA.

Fronds bright crimson, flat, leaf-like, membranaceous, transversely veined, traversed by a strong midrib. Spores egg-shaped, contained in spherical conceptacles, which are borne on short stalks, mostly on one side of the midrib; tetraspores numerous, produced in winter in pod-like leaflets, which grow on the midribs of fronds which have lost their membrane.

The beautiful plant, which is at present the only representative of this genus, is an old favourite, appearing under a new name. It is nearly the largest, and by far the most striking and brilliant British sea-weed of the Red series, and is familiar to every collector as *Delesseria sanguinea*. Unfortunately this name could not be retained, for under the classification of Agardh, which is now generally adopted, it was necessary to remove the plant out of the genus *Delesseria*. Dr. Harvey, in consequence, fell back on the name *Wormskioldia*, which was published by Sprengel, in 1827; but my friend Mr. Carruthers has pointed out that this name was previously appropriated, in De Candolle's 'Prodromus,' in 1824, to a genus of flowering plants, and that it is not, therefore, available. In these circumstances, it occurred to me that this handsome plant would be a fitting monument to dedicate to the memory of Dr. Harvey, and accordingly the name *Harveya* was inserted in my MS. I subsequently found that Dr. Harvey's labours had been already recognized, and I was therefore prevented the pleasure of offering this tribute of respect and esteem to my kind and distinguished friend. I have sought for some worthy recipient of algological honours, and failing to find a candidate known to fame whose name is not already appropriated, I venture, with all diffidence, to call the genus *Maugeria*, in honour of Mrs. W. P. Mauger, a very accomplished and diligent student of British Sea-weeds, to whom I am indebted for much valuable assistance in the preparation of this work, and for the use of many rare, and some almost unique specimens.

Maugeria sanguinea. The crimson Maugeria.

Frond a cylindrical, branched stem, beset throughout its length with numerous, irregularly-placed leaves; leaves from a few inches to a foot or more in length, and from half an inch to two or three inches in width, transversely veined, entire at the margin, when young very delicate, and perfectly flat, becoming a little coarser and more or less torn with age. Fructification, of both kinds, developed during winter on the midribs of old leaves, the membrane of which has decayed; spores in short-stalked, spherical conceptacles; tetraspores closely massed in minute leaves, called *sporophylla*.

This species is perennial, and is very widely distributed, being found on the Atlantic shores of Europe, in the Baltic Sea, and again in various localities in the Southern hemisphere. It is more or less abundant all round our coasts, and in favourable situations grows to a large size. It is to be found, in deep tide-pools, or on the perpendicular sides of rocks just beyond low-water mark. The fronds begin to grow in early spring, and soon attain their full beauty. The best specimens, that is to say, those with perfect leaves and of brightest colour, are to be obtained not later than the end of June; but the spores and tetraspores are only formed in winter. It may appear at first sight an unnatural arrangement to separate this species from the *Delesseriæ*, with which it was formerly associated; but the differences of the fructification are so important, that it is necessary to remove it, not only to another genus, but to a different Order.

Genus LXXI. RHODYMENIA.

Frond flat, membranaceous, forked or hand-shaped, often proliferous from the margin, composed of two series of cells; those within oblong, those on the surface minute. Spore-threads very numerous, issuing from a basal placenta and forming a simple nucleus in sessile, hemispherical conceptacles scattered over the frond, and opening at length by a terminal pore; tetraspores roundish, cruciate or tripartite, either collected in cloud-like patches, or dispersed among the surface-cells of the frond.—RHODYMENIA, from the Greek *rhodeos*, rosy, and *umeen*, a membrane.

This genus formerly included a large number of plants, which resembled each other in habit and other respects, but which more accurate observation has shown to be very widely separated by difference of structure. Professor Agardh has reduced the species to about a dozen at most, and only two of these are British; even this curtailed genus is divided into two sections, each of which contains one of our species.

Rhodymenia palmata. The hand-shaped Rhodymenia.

Fronds from a few inches to more than a foot long, and often nearly as wide, wedge-shaped at the base, irregularly cleft in a forked or palmate manner; the margin generally flat, but sometimes furnished with small leaflets, which give the frond a pinnate appearance; substance membranaceous, becoming tough and leathery with age. Root a small disc. Tetraspores distributed over the whole frond, in cloud-like spots.

This plant assumes so many varied forms, either from peculiar conditions of growth, from age, or from mere caprice, that it would be wasted labour to attempt to describe them all. They are best studied in Nature's boundless schoolroom among the rocks, and in order to impress them on the mind it is desirable to collect as complete and varied a series of specimens as circumstances may permit. The following forms are, perhaps, worthy of brief notice:—

Var. β. *marginifera*. Frond with leaflets on its margin.

Var. γ. *simplex*. Frond undivided, wedge-shaped.

Var. δ. *Sarniensis*. Frond divided into linear segments.

Var. ε. *sobolifera*. Frond with a stem, and very narrow, much divided branches, which expand into wedge-shaped, jagged lobes, and become fringed at the tip.

The last is the only form which I have observed to be very constant.

Rhodymenia palmetta. The little Palm Rhodymenia.

Fronds red, fan-shaped, more or less forkedly cleft, expanding from a cylindrical stem into linear, wedge-shaped segments, with broad, rounded interstices, and a flat margin. Spores angular, contained in stalkless conceptacles, which are either marginal or scattered over the frond; tetraspores cruciate, forming deep red patches on the tips of the segments.

This little plant grows on rocks, shells, and the stems of large sea-weeds, in deep water. It is less common than *R. palmata*, but is found in many localities. Specimens bearing spores differ considerably from those with tetraspores. In the former the frond is small, with short, crowded segments, and is borne on a long, simple stem; while in the latter it is spreading, and the stem branches almost from the base. There is a variety of this species, with a simple or once-forked, narrow frond, rising from a fibrous root, which Professor Agardh calls *Nicæensis*. It is not common, and resembles *Phyllophora Brodiæi*.

Genus LXXII. EUTHORA.

Frond flat, membranaceous, forkedly-pinnate, composed of two series of cells; those within large, oblong, those on the surface coloured, minute. Spore-threads very numerous, radiating from a central placenta, and attached to the walls of the conceptacle by subsimple threads; tetraspores cruciate, lodged in the thickened tips of the frond.—EU-THORA, from the Greek *eu*, well, and *thoros*, a seed.

This genus was formerly combined with *Rhodymenia*, and has been separated on account of a difference in the arrangement of the spore-threads, which radiate from a central placenta, instead of rising from a basal one, as in that genus. These characters can only be seen by dissecting a spore-conceptacle under a microscope, and this is not by any means an easy operation to perform on so minute a subject.

Euthora cristata. The crested Euthora.

Frond fan-shaped, membranaceous, one to three inches high, divided into numerous segments, which expand upwards, and are repeatedly subdivided, the secondary segments being alternate, linear, jagged at the tips, and often fringed. Spores in globular conceptacles, which are arranged irregularly along the margin of all the divisions of the frond; tetraspores cruciate, in the ultimate branchlets. The branches of those plants which bear tetraspores are more slender than those of plants with tubercles.

This species is very rare, and is only found on our northern shores. It is of a beautiful bright red colour, and varies much in the width and division of the frond. It grows on the stems of *Laminariæ*, in deep water, and is a summer annual.

Genus LXXIII. **RHODOPHYLLIS.**

Frond flat, membranaceous, forkedly cleft, often lobed on the margin, composed of two series of cells; those of the inner series large, longitudinal, those of the outer vertical, in few rows. Spore-threads very numerous, radiating from a basal placenta and forming a compound nucleus in marginal, nearly spherical, closed conceptacles; tetraspores zonate, immersed in the frond, or in marginal lobes.—RHODOPHYLLIS, from the Greek *rhodeos*, rosy, and *phyllon*, a leaf.

This is another division of the old genus *Rhodymenia*, consequent on the new arrangement, according to fructification. It differs from the reformed genus *Rhodymenia*, in the structure of the spore-nucleus, and in having zonate tetraspores.

Rhodophyllis bifida. The cleft Rhodophyllis.

Fronds growing in tufts, from one to four inches high, transparent, without veins, forkedly divided from the base, of a brilliant carmine colour. Spores in globose, stalkless conceptacles, which are usually numerous on the margin of the frond; tetraspores oblong, marked with three transverse lines, arranged in patches in the upper segments.

This is a very variable plant, and is unfortunately so rare that it is difficult to obtain a series of the different forms. It grows on rocks, etc., in deep water, and is sparingly distributed round the coasts of Great Britain, Ireland, and the Channel Islands. It is a summer annual. Dr. Harvey describes two varieties. One of these, var. β. *ciliata*, has since been raised to the rank of a species, and is described below. The other is "var. γ. *incrassata*. Frond thicker than usual, shrinking, and turning to brownish-red in drying, broad; segments cruciate, proliferous, or ciliate at the margin."

Rhodophyllis appendiculata. The ciliated Rhodophyllis.

Fronds growing in tufts, from one to two inches high, narrower and thicker than those of *R. bifida*, much divided, opake, of a brownish-red colour, their edges fringed with leafy cilia, which contain tetraspores.

This plant has but recently been raised to the rank of a species. It was formerly considered to be a variety of *R. bifida*, and is figured and described in the 'Phycologia Britannica' as *R. bifida*, var. β. *ciliata*. It is very rare.

Genus LXXIV. PLOCAMIUM.

Frond somewhat cartilaginous, linear, flattened, pinnately divided, composed of two strata of cells; the inner longitudinal and oblong, the outer many-sided and small. Spores on filaments radiating in tufts from a basal placenta in hemispherical conceptacles, which are either with or without stalks; tetraspores zonate, oblong, in small leaf-like stichidia.—PLOCAMIUM, from the Greek *plokamos*, braided hair.

This is a very beautiful and widely distributed genus, all the species of which are remarkable for their brilliant colour and handsome tree-like fronds.

Plocamium coccineum. Scarlet Plocamium.

Frond cartilaginous, narrow, flattened, much divided; branches spreading, irregularly alternate; branchlets awl-shaped, curved, set with comb-like spurs on their inner side. Spores in solitary, marginal, stalkless conceptacles; tetraspores zonate, in leaf-like stichidia on the inner sides of the ramuli.

This is one of the most abundant, and is probably the

best known of the Red division of the British sea-weeds. The facility with which its flattened branches lay themselves out and adhere to paper, its elegant tree-like form and brilliant lasting colour, have established it as the prime favourite of unlearned sea-weed collectors, and makers of sea-weed ornaments. It is generally distributed all round our coasts, and, indeed, throughout the temperate zone. It varies considerably in size and substance, and in the form of its branches, so much so that some of the more slender specimens are liable to be considered a distinct species. Its general characters are, however, sufficiently constant to be easily determined by a careful examination.

Genus LXXV. CORDYLECLADIA.

Frond thread-like, irregularly branched, cartilaginous, formed of two series of cells, those in the centre oblong, longitudinal; those on the surface roundish, minute, vertical. Spores roundish, formed on branched threads, arranged in a dense globular mass in spherical conceptacles without stalks; tetraspores oblong, cruciate, immersed in pod-like branchlets.—CORDYLECLADIA, from the Greek *chorde*, a string, and *klados*, a shoot.

The sole representative of this genus in the British flora—and it is doubtful if any other species exist elsewhere—was formerly included in the genus *Gracilaria*, whence it has been removed by Professor Agardh, for reasons already referred to in speaking of the Order *Rhodymeniaceæ*.

Cordylecladia erecta. The erect Cordylecladia.

Fronds two or three inches high, cylindrical, thread-like,

rising in groups from a common disc-root, irregularly branched, sometimes simply forked at the top, and sometimes having a few side branches. Spores massed in the centre of thick spherical conceptacles, which are densely clustered on the branches; tetraspores oblong, cruciately divided, contained in lance-shaped pods at the tips of the branches.

This interesting little plant grows at the bottom of sandy pools, and fruits in winter, when it may be easily recognised. In a barren state it closely resembles a stunted growth of *Gracilaria confervoides*, and this fact may probably account, in part, for the small number of recorded specimens, as the plant is no doubt frequently overlooked in summer and autumn, which are usually the most convenient seasons for collecting.

Order XVIII. CRYPTONEMIACEÆ.

*Purple or Rose-red Sea-weeds with an inarticulate, horny, cartilaginous, gelatinous, and rarely coriaceous or membranous frond, composed of confervoid filaments set in gelatine. The membranous species are sometimes composed of many-sided cells, gradually decreasing in size towards the surface. Fructification:—*1. *Spores congregated without order in cells sunk in the frond, or in external conceptacles.* 2. *Zonate or cruciate tetraspores dispersed among the cells of the periphery, collected in definite sori, or, more rarely, in wart-like bodies, called nemathecia.*

Genus LXXVI. STENOGRAMME.

Frond flat, dichotomous, proliferous from the margin, composed of two strata of cells; those of the inner stratum many-sided and empty, those on the surface minute and coloured. Spores roundish, in dense clusters, disposed in linear conceptacles, which traverse the middle of the frond; tetraspores in wart-like superficial conceptacles.—STENO-GRAMME, from the Greek *stenos*, narrow, and *gramma*, a line.

The single species composing this genus grows in both hemispheres, though it has never been found in great abundance on any coast. It was first discovered in Spain, in 1823, and has since been dredged in the south of England and in Cork Harbour, but the largest specimens have been brought from California and New Zealand.

Stenogramme interrupta. The interrupted Stenogramme.

Frond of a clear pink colour, on a short stem, rising from a disc-root, and rapidly expanding into a fan-shaped membrane, varying from three to five inches in breadth; some specimens are deeply cleft into ribbon-like laciniæ, others are less divided, and occasionally throw out oblong or forked leaflets. Spore-conceptacles contained in a nerve running through the centre of the fertile lobes; tetraspores in dark-red warts, called nemathecia, scattered irregularly over both surfaces of the frond.

This is one of the most rare of our native sea-weeds, and is the more difficult to obtain as it grows in deep water, and is seldom cast on shore. At first sight the frond appears to be traversed by a midrib, but on closer examination it will be found that the peculiar form and arrangement of the spore-conceptacles give it this appearance, and that barren fronds and those which bear tetraspores are wholly nerveless. I am indebted to Mr. John Gatcombe, of Plymouth, for specimens of this valuable plant, which he collected in the locality, where it was originally found by Dr. Cocks.

Genus LXXVII. PHYLLOPHORA.

Frond stalked, the stalk expanding into a rigid, membranaceous, flat lamina, proliferous from the disc or margin, without veins, or slightly veined at the base. Spores minute, contained in sessile or stalked conceptacles; tetraspores cruciate, contained in external, scattered warts.— PHYLLOPHORA, from the Greek *phyllon*, a leaf, and *phero*, to carry.

The plants of this genus are mostly common on exposed coasts; they grow near low-water mark, or at a greater depth. Some of them are to be found on the Atlantic coasts of both Europe and America, and some extend into the Baltic Sea; others are content with a more limited range.

Phyllophora Brodiæi. Brodie's Phyllophora.

Frond of a deep red colour, composed of a cylindrical, branched stem, three to four inches or more long, the branches expanding into oblong or wedge-shaped lobes, which are frequently proliferous from the extremities. Root a small disc. Spore-conceptacles globose, stalkless, developed on the lamina; tetraspores in warts, at the tips of the frond.

This plant is one of the least common of the genus *Phyllophora*, and is chiefly confined to our northern shores. It adheres very imperfectly to paper in drying. It is said to be perennial, and grows on rocks in deep water during winter and spring. Dr. Harvey describes a "var. β. *simplex*. Stem short, expanding into an oblong, simple or once-forked, rose-coloured frond; sorus elliptical, composed of tetraspores."

Phyllophora rubens. The red Phyllophora.

Fronds densely tufted, expanding from a short, cylindrical stem, into very blunt, wedge-shaped segments, either single or forked, varying from a quarter of an inch to half an inch in width; a second series of similar segments springs from the tips of the first, and this process is repeated as the plants advance in age; all the segments are minutely stalked, and have an indistinct midrib. Spores minute, in spherical conceptacles, which are either scattered over the surface of the frond, or arranged in a line within the margin; tetraspores unknown.

This handsome species is common on most of our shores, but is rarely seen in its full beauty, as it is almost always covered with minute shells and *Melobesiæ*. These interfere with the process of drying, which is already sufficiently difficult, owing to the rigid nature of the plant. It grows in deep tide-pools, under the shelter of the large *Laminariæ*, and likewise on rocks, etc., in water of a considerable depth. It is perennial, and in perfection in winter.

Phyllophora membranifolia. The membrane-leaved Phyllophora.

Fronds from three to twelve inches in height, growing in tufts from an expanded root; stem cylindrical, branched, expanding into broadly wedge-shaped, forked laminæ. Spores in large, stalked conceptacles, which rise from the branches or the laminæ; tetraspores in wart-like bodies, forming dark-coloured convex patches in the centre of the laminæ.

The longer stem and purple colour of this plant readily distinguish it from the other species of the genus. This species is very common on our coasts. It is perennial, and grows on rocks, etc., between the tide-marks.

Phyllophora palmettoides. The palmetto-like Phyllophora.

Root a widely-expanded disc; stem cylindrical; frond oblong, simple or forked, proliferous from the tips, of a beautiful rose-red colour. Spores unknown; tetraspores cruciate or tripartite, in solitary sori, which are immersed in the substance of the frond, near the tip.

This beautiful little plant, which was formerly considered to be a variety of *Phyllophora Brodiœi*, is found only on our southern shores. It is distinguished by the bright colour of its frond, its more widely-expanded root, and the position of the tetraspores. It grows on rocks near low-water mark in winter and spring, and is perennial.

Genus LXXVIII. **GYMNOGONGRUS.**

Frond tapering, compressed or flat, linear, forked, the centre cells roundish, those of the surface minute, arranged in closely-packed, vertical, necklace-like filaments. Spores minute, arranged in masses (nucleoli), several of which are associated together into a conceptacle-like cluster, which is immersed in the frond; tetraspores cruciate, formed in necklace-like threads, in external, hemispherical, wart-like excrescences (nemathecia).—GYMNOGONGRUS, from the Greek *gymnos*, naked, and *goggros*, an excrescence on a tree.

This genus comprises at present some twenty species, which are widely distributed, but two only are natives of Great Britain.

Gymnogongrus Griffithsiæ. Mrs. Griffiths's Gymnogongrus.

Root an expanded disc; fronds densely tufted, one to three inches high, simple at base, then repeatedly forked; branches flexuous and tapering, except at the tips, which are sometimes slightly flattened. Tetraspores cruciate, elliptical, formed of the upper joints of necklace-like filaments, massed together in warts, either on the side of the frond, or encircling it.

This plant is perennial, and grows between the tide-marks. It is in perfection in autumn and winter, when it is generally thickly studded with nemathecia. These should be examined under a microscope, for they are extremely beautiful. The tetraspores are of a delicate pink colour, transparent and sparkling, and are arranged in chains, which a slight effort of the imagination will easily convert into necklaces of tiny rubies.

Gymnogongrus Norvegicus. The Norwegian Gymnogongrus.

Stem short, cylindrical; fronds linear, flat, repeatedly forked, two to three inches or more high, more or less tufted; the axils of the branches spreading, rounded, the tips blunt. Spores very numerous, in minute, depressed, spherical conceptacles, immersed in the upper segments of the frond, and projecting on both sides; tetraspores in stalkless nemathecia, which are thickly scattered over both sides of the frond.

This species was formerly known as *Chondrus Norvegicus*, but has been removed to its present position by Professor Agardh. It is not very common, nor are its characters easily recognized. Although called *Norve-*

gicus, it is by no means peculiar to Norway. It grows on rocks, near low-water mark, during winter and spring, and is annual.

Genus LXXIX. AHNFELTIA.

Frond horny, tapering, forked, composed of two strata of cells; those in the centre very slender, elongated, densely packed; those on the surface minute, arranged in vertical, closely packed, short, necklace-like filaments. Spores minute, arranged in masses (nucleoli) several of which are associated together into a conceptacle-like cluster, which is immersed in the frond; tetraspores in external wart-like excrescences surrounding the branches.

The only British species of this genus was formerly included in *Gymnogongrus*, from which it has been separated on account of the much greater density of cellular structure, and more rigid substance of the frond.

Ahnfeltia plicata. The entangled Ahnfeltia.

Fronds from six to ten inches or more long, horny, taper, thread-like, irregularly branched, tangled; axils rounded; terminal branches blunt at the tip. Fructification in wart-like excrescences scattered over the branches: spores and tetraspores being but seldom fully developed.

This is a peculiar plant, and cannot easily be mistaken for any other species. It is widely distributed both in this country and elsewhere; it is perennial, and grows on rocks and stones at various depths; the fronds are usually tangled together into a dense mass, they are of a rigid, wire-like texture, of a dark purple colour, and about as thick as fine twine.

Genus LXXX. CYSTOCLONIUM.

Frond fleshy, taper, much branched, composed of three kinds of cells; those in the centre cord-like, formed of elongated longitudinal fibres; the second series large and round; the external, or bark layer, small and angular. Spores minute, arranged in masses, several of which are enclosed in a thick pericarp, in conceptacles which are partly immersed in the branches; tetraspores zonate, scattered among the external cells of the branches.—CYSTOCLONIUM, from the Greek *kustis*, a bladder, and *kloon*, a young shoot.

The present position of this genus is very different from that which it formerly occupied, and has been assigned to it in consequence of the arrangement of the fructification. The structure of the frond is, however, the character by which the plants composing the genus are distinguished from their immediate allies.

Cystoclonium purpurascens. The purple Cystoclonium.

Frond taper, much branched; branches alternate, elongate; branchlets numerous, tapering to each end. Spores in conceptacles embedded in the branchlets, either singly or two or more together; tetraspores zonate, immersed among the surface cells of the branchlets.

This species is a summer annual, and grows on rocks, etc., between the tide-marks. It is one of the most common of our native sea-weeds, but is best known under its old name of *Hypnea*. There is one well-marked variety, β. *cirrhosa*. It is irregularly branched and variously distorted; the branches are zig-zag, here and there swollen, their apices are lengthened into tendrils, which coil round the stems of neighbouring plants.

Genus LXXXI. CALLOPHYLLIS.

Frond fleshy, membranaceous, blood-red, flat, forked, formed of two strata of cells; the inner stratum of large, roundish cells, each of which is surrounded by a network of cellules; the outer, or cortical stratum of vertical, necklace-like filaments. Spores minute, arranged in masses, several of which are enclosed in a thick pericarp, in conceptacles immersed in the margin; tetraspores cruciate, immersed in the frond.—CALLOPHYLLIS, from the Greek *kalos*, beautiful, and *phyllon*, a leaf.

The plants which compose this genus were formerly included among the *Rhodymeniæ*, which they much resemble in external appearance, but from which they differ in the structure of the frond, and the arrangement of the spores. They are chiefly natives of warm climates, and are remarkable for their handsome fronds, which are usually of a bright red colour.

Callophyllis laciniata. The jagged Callophyllis.

Frond fleshy, of a bright red colour, fan-shaped or sometimes palmate, cleft into numerous, broad, wedge-shaped segments, which are again forkedly divided; the margins are sometimes proliferous, and those of fertile specimens are curled, and fringed with minute leaflets. Spores minute, in conceptacles lodged in the marginal leaflets; tetraspores tripartite or cruciate, arranged in cloudy patches in the substance of the frond.

This species is subject to many variations of form, but is always conspicuous for its brilliant colour. It is found on most parts of our coast, chiefly in bold rocky localities where it grows, generally in deep water, on rocks and *Laminariæ*. It is said to be biennial, and is in perfection during spring and summer. It was formerly called *Rhodymenia laciniata*.

Genus LXXXII. KALLYMENIA.

Frond fleshy, membranaceous, flat, expanded, with an irregular outline, composed of three strata; the centre consisting of longitudinal, branching, interlaced, articulated filaments; the second of large, many-sided cells, and the third, or cortical, stratum of minute cells. Spores minute, arranged in masses which are enclosed, several together in an indistinct pericarp, in conceptacles immersed in and projecting on both sides of the frond; tetraspores cruciate, scattered among the surface cellules.—KALLYMENIA, from the Greek *kalos*, beautiful, and *umeen*, membrane.

There are two British species of this genus, both of which are rare.

Kallymenia reniformis. The kidney-shaped Kallymenia.

Stem short; frond when young kidney-shaped or roundish, becoming irregularly cleft and lobed with age, sometimes producing young fronds on its margin. Spores in conceptacles, which are densely scattered over, and half immersed in the frond; tetraspores minute.

The recorded habitats of this species extend from Orkney to the Channel Islands; it is nowhere abundant, but more so in the south than in the north. It grows in deep, shady pools at extreme low-water mark, is perennial, and in perfection in summer and autumn.

Kallymenia microphylla. The small-leaved Kallymenia.

Frond "kidney-shaped, broadly expanded; conceptacles emerging from one side only, nearly flat above."

I have extracted the description of this species from Dr. Gray's 'Handbook of British Water-weeds,' where it is inserted on the authority of Professor Agardh. I have never seen a specimen, nor am I aware that the species has been included in any previous work on British Algæ.

Genus LXXXIII. **GIGARTINA.**

Frond cartilaginous, flat or cylindrical, simple or branched, composed of cylindrical, jointed fibres, arranged in a loose network, surrounded by a bark formed of necklace-like, forked filaments, set in gelatine. Spores roundish, arranged in confluent masses, in a pseudo-pericarp, formed of closely interwoven filaments, in external, globose conceptacles, which are furnished with a terminal pore; tetraspores cruciate, arranged in somewhat prominent masses, beneath the surface cells.—GIGARTINA, from the Greek *gigarton*, a grape seed.

This genus is nearly allied to *Chondrus,* which it resembles in general character. The *Gigartinæ* may be distinguished by their external conceptacles, and by the pseudo-pericarp in which their spores are enclosed.

Gigartina acicularis. The needle-branched Gigartina.

Root disc-like, with branching fibres; fronds two to four inches high, as thick as small twine, irregularly branched; branches short and spine-like, or lengthened, and furnished with a second series, always acutely pointed. Spores in spherical conceptacles, which protrude from the branches and branchlets.

This plant grows on rocks near low-water mark, and is annual. It occurs most frequently on the south and west of England, and on the shores of Ireland and the Channel Islands. It is considered rare, but this may probably be partly due to the fact that it matures in winter, when it is very likely to be overlooked. Perhaps the best way to make out the genus is to examine a thin, vertical section of a frond under a microscope, which will reveal the peculiar arrangement of netted longitudinal fibres in the centre, and the closely-packed, vertical filaments which form the bark. The specific distinction is less difficult to establish, as the general habit of the plant is very different from that of either of the other British species.

Gigartina pistillata. The pedicellate Gigartina.

Root a broad disc, without fibres; frond two to six inches high, flattened, destitute of branches below, branched in a fan-like manner above; branches repeatedly forked, spreading, with rounded axils and acute tips. Spores in globular conceptacles, which are produced abundantly on the branches, either singly or two or more together; tetraspores cruciate, immersed in clusters in the substance of the branches.

This species is even more rare than the last; it appears to be most abundant on the coasts of Cornwall and Devon. The mode of branching, and the form and greater number of the conceptacles are very obvious and well-defined characters. This plant grows on rocks, near low-water mark, in winter, and is perennial.

Gigartina Teedii. Teed's Gigartina.

Fronds from three to six inches long, growing several together from the same base, much branched; stem flat, about a quarter of an inch broad in the middle, tapering at each end; branches opposite or alternate, once or twice pinnate, closely beset with short, spine-like branchlets. Spores in globose, stalkless conceptacles, seated on the branches; tetraspores in roundish sori, near the margin of the frond.

This is the most rare of all our native sea-weeds. The habitat in Elberry Cove, Torbay, where it was first discovered by Mrs. Griffiths, is the only one in this country that I am aware has been recorded. On the coast of Normandy it is more abundant, and it is common in the Mediterranean. It would thus appear to have reached its northernmost limit on our shores. It grows on rocks, at or a little beyond extreme low-water mark, and is perennial. When fresh, the plant is of a deep red colour, which in decay changes to a bright green. In drying, the frond shrinks considerably, and does not adhere to paper.

Gigartina mamillosa. The mamillose Gigartina.

Stem linear, channelled; fronds about six inches high, growing in tufts, fan-shaped, divided into wedge-shaped, cleft segments, with incurved margins, or sometimes nearly linear throughout. Spores in ovate conceptacles, lodged in mamilliform or filiform processes, which are thickly studded over the surface of the frond.

Most localities on our coast produce this species, which grows near low-water mark in winter and spring, and is perennial. When in fruit, the characters are sufficiently distinct; but barren or narrow fertile specimens much resemble *Chondrus crispus*. They differ, however, in having their stems channelled, and the margin of their fronds incurved.

Genus LXXXIV. CHONDRUS.

Frond cartilaginous, flat, forked, composed of two kinds of cells; those in the centre forming a network of cylindrical, jointed fibres; those of the bark necklace-like, arranged vertically in gelatine. Spores minute, in confluent masses, several together forming a roundish nucleus, which is immersed in the substance of the frond, without definite border; tetraspores cruciate, in sori, lodged beneath the surface cells.—CHONDRUS, from the Greek *chondros*, cartilage.

This is a small genus, but is widely distributed, and very abundant.

Chondrus crispus. The curled Chondrus, Carrageen, or Irish Moss.

Frond stalked, fan-shaped, forked, flat, the segments of very variable shape, from linear to broadly-cuneate. Spores in oval clusters, immersed in the frond, prominent on one surface, depressed on the other; tetraspores in small, red sori, scattered over the frond.

In brackish water, and in still, deep pools, this species attains a very broad, luxuriant growth, and sometimes becomes curled and fringed at the margin; in exposed situations, it is much smaller and more narrow. It is very common, but its numerous varieties are extremely interesting, and may be illustrated by an extended series of specimens. It is the Carrageen or Irish moss of commerce, and when boiled produces a clear, tasteless gelatine, which is occasionally used to make blancmange, and as a remedy for consumption. This plant is perennial, and may be found during spring and summer on every rocky shore. It grows between the tide-marks and in deep water.

Genus LXXXV. CHYLOCLADIA.

Frond tubular, rounded or somewhat flattened, much branched; its inner stratum composed of elongated and anastomosing filaments; its outer stratum of roundish, polygonal cells, which become smaller towards the surface, and form a membranaceous bark. Spores formed on fibres, which radiate from a placenta, enclosed in a transparent sac, in external, conical conceptacles with a terminal pore; tetraspores tripartite, scattered among the surface cells of the branches. —CHYLOCLADIA, from the Greek *chylos*, juice, and *klados*, a shoot.

The species composing this genus have been changed under the new arrangement. Those which were formerly *Chylocladiæ* are now, with one exception, *Lomentariæ*, and the *Chrysymeniæ* have become *Chylocladiæ*.

Chylocladia articulata. The jointed Chylocladia.

Fronds growing in dense tufts, from two to ten inches long. tubular, filled with fluid, somewhat gelatinous, strongly constricted throughout, much branched; the lower branches forked, the upper pinnate, whorled, tufted. Spores in obtusely conical conceptacles, which have thick walls composed of small cells, and a minute terminal pore, and are scattered over the upper joints of the frond; tetraspores tripartite, lodged in the tissue of the joints.

This is the only British species of the old genus *Chylocladia*, which retains its name. It is common all round our coast during summer, and is annual. It grows on rocks, between the tide-marks, usually under

Fuci and other large weeds. It resembles the *Opuntia*, or prickly-pear, in habit, and the whole plant is made up of a number of similar joints springing one from another, at first singly, then three together, or whorled.

Chylocladia clavellosa. The clubbed Chylocladia.

Fronds from a few inches to a foot long, gelatinously membranaceous, much branched; branches pinnate, crowded, thickly set with short, spindle-shaped branchlets. Spores angular, in ovate or conical conceptacles; tetraspores immersed in the branchlets.

This species is moderately abundant on all parts of the British coast, from the extreme north of Scotland to the Channel Islands. It is a well-marked, handsome plant, of tender substance and light red colour, which becomes darker in drying. It is annual, and grows near low-water mark, or at a greater depth, either on rocks or on the stems of *Laminariæ* and other Algæ.

Chylocladia rosea. The rose-coloured Chylocladia.

Fronds one to two inches high, pinnately branched, hollow, all the divisions when young broadly spindle-shaped, lengthening with age; the branchlets ultimately becoming long and very narrow. Tetraspores tripartite, scattered in irregular patches over the frond.

Mr. Gatcombe writes from Plymouth, "I have invariably found this species growing on the sheltered sides of perpendicular rocks, in pools, or when left dry at very low water, and I feel confident that owing to the great resemblance it bears to young *Delesseria Hypoglossum*, it must have been frequently overlooked by algologists." And again, in another letter, received a fortnight later, he writes:—" On Friday I succeeded, after wading up to my waist, in reaching the rock on which *Chrysymenia (Chylocladia) rosea* grew, and found the plants so altered in appearance that I hardly knew them; they had become so spiry, and altogether different in form and colour, that they did not deserve the name of *rosea* at all. However, you will now be able to describe the full-grown plants from the specimens I intend to forward in a few days." *C. rosea* is said to be annual, and to grow on rocks and Algæ in deep water, and it is evident from the above account that it likewise grows between the tide-marks. Orkney and Filey are recorded as habitats by Dr. Harvey, and to these Plymouth must be added.

Genus LXXXVI. HALYMENIA.

Frond cylindrical or flat, gelatinous or fleshy, forked or pinnate, consisting of a thin double membrane, composed of small, coloured cells, and separated internally by jointed, branching fibres. Spores minute, densely packed, enclosed in a transparent sac, which is immersed in the frond immediately beneath the surface; tetraspores cruciate, and scattered among the surface cellules.—HALYMENIA, from the Greek *als*, the sea, and *umeen*, a membrane.

The specimens included in this genus are all of a beautiful rose-colour, and delicate gelatinous substance; only one is found on our coasts, and that but sparingly.

Halymenia ligulata. The strap-shaped Halymenia.

Root a small shield; frond from a few inches to a foot or more long, and from an eighth of an inch to two inches broad, very variably divided, sometimes quite simple, at others many times forked and proliferous, always gelatinous. Spores in minute dots, abundantly scattered over the whole frond.

Although subject to extreme variation of size and form, this is not a difficult species to determine. The gelatinous texture of the frond, taken in conjunction with its general character, always affords a ready means of identification. I have collected many specimens in Jersey, where the plant is moderately abundant in summer. It is annual, and grows on rocks near low-water mark, and at greater depths, whence it is cast on shore. Specimens dry easily, and do not require to be much pressed.

Genus LXXXVII. FURCELLARIA.

Frond, cylindrical, forked, solid, composed of three kinds of cells; those in the centre longitudinal, elongate, interwoven; those of the next series roundish, large; and those of the outer, or bark series, small, arranged vertically in necklace-like threads. Spores large, angular, arranged in nuclei (favellæ), which are immersed several together, in the podlike tips of the branches; tetraspores large, pear-shaped, zonate, formed just within the external layer of cells, from the outer fibres of the intermediate layer.—FURCELLARIA, from the Latin *furcula*, a little fork.

It is a remarkable result of the new system of classification, based on the mode of fructification, that two genera so much alike in their general characters as *Polyides* and *Furcellaria* should be so widely separated.

Furcellaria fastigiata. The pointed Furcellaria.

Root fibrous; fronds from a few inches to about a foot high, cylindrical, repeatedly forked; branches all of the same height, with acute axils and tips. Spore-clusters formed from and embedded among the intermediate cells of terminal, pod-like receptacles; tetraspores zonate, scattered among the external cells of similar receptacles.

This species is very common. It grows on rocks in tidal pools, is perennial, and fruits in winter. The pods, which are formed at the tips of the branches, and contain the spores and tetraspores, are very peculiar. Their structure is similar to that of the rest of the frond, but they are separated from it, as though by a joint, and fall off when the fruit is mature, leaving the branches truncated. In external appearance and structure this species resembles *Polyides rotundus* so closely as to be with difficulty distinguished. The fructification is however very different, and the root of the one is fibrous, and of the other a disc; but this latter character is not always easy to make out, or quite trustworthy.

Genus LXXXVIII. GRATELOUPIA.

Frond compressed or flat, membranaceous, pinnate, consisting of two series of thread-like cells; those of the inner series densely interwoven, slightly jointed; those of the outer, or bark series short, necklace-like, horizontal. Spores numerous, enveloped in a transparent, gelatinous membrane, and immersed, in simple nuclei (favellæ), just beneath the external series of cells: these nuclei project slightly beyond the surface of the frond, and are generally developed in groups; when mature, the spores are discharged through a pore in the surface of the frond; tetraspores cruciate, immersed among the surface-cells, scattered.—GRATELOUPIA, in honour of Dr. Grateloup, a French naturalist.

The majority of the species comprised in this genus grow in warm latitudes, and only one is found as far north as our southern coast. One species, *G. Gibbesii*, is the largest sea-weed yet discovered in Charleston Harbour, where its fronds attain a length of nearly two feet, and a width of about an inch and a half. The contrast between this plant and our own pigmy *G. filicina*, which is seldom more than three or four inches high and an eighth of an inch broad, is very striking.

Grateloupia filicina.   The fern-like Grateloupia.

Frond somewhat flattened, twice or thrice pinnate; pinnæ linear, narrow at the base, pointed at the tip, longest at the lower part of the frond, which has in consequence a pyramidal outline. Spore-nuclei immersed in the branches; tetraspores cruciate, scattered among the surface cells of accessory leaflets.

The only plant with which this species is likely to be confounded is one of the numerous forms of *Gelidium corneum*, but the structure of the frond, and the arrangement of the spore-nuclei and tetraspores are very different. In the autumn of 1865 I found numerous specimens of this rare plant in both kinds of fruit in shallow rock-pools, near high-water mark, in St. Brelade's Bay, Jersey. "Submarine rocks, about half-tide level, frequently where small streamlets run into the sea," are the usual habitats of this species, which is perennial, and attains its most perfect state in autumn and winter.

Genus LXXXIX. SCHIZYMENIA.

Frond flat, fleshy, composed of two layers; the inner of densely interwoven, slightly-jointed threads; the outer of vertical, necklace-like threads. Spore-nuclei immersed in the outer layer of the frond; tetraspores cruciate, collected into dense sori, also lodged in the outer layer, but in separate plants.—SCHIZYMENIA, from the Greek *schistos*, cloven; and *umeen*, a membrane.

There are about a dozen species described as belonging to this genus, but only two of them are British, and these have been hitherto known to collectors by different and separate generic names. One was formerly the well-known *Iridæa edulis*, and the other, the less common *Kallymenia Dubyi*.

Schizymenia edulis. The eatable Schizymenia.

Fronds growing in clusters, several from the same base, and varying in size from a few inches to a foot or more in length, and from two to six inches broad, commencing with a short, cylindrical stem, which becomes gradually flat, and expands into an oval, thick, leathery, leaf-like disc. Spore-clusters sunk beneath the external layer of cells in the upper part of the frond, and having the appearance of minute dots; tetraspores cruciate, lodged in a dense, band-like sorus, just within the periphery.

This plant, the *Iridæa edulis* of the 'Phycologia Britannica' and other works, seems to have been destined to a plurality of names throughout its career. Dr. Harvey mentions no less than six generic, and four specific names for it, and these do not include that of *Schizymenia*, its present, and I trust its permanent designation. Add to this, that it is called *edulis*, without being eatable, and the 'Comedy of Errors' may be considered to be complete. It grows near low-water mark, is perennial, and fruits in winter. It occurs on all parts of the British coast, from Orkney to the Channel Islands, and indeed extends far beyond these limits, both north and south. The only variation to which it is liable is that of size, but the fronds are frequently much torn and altered in appearance by the action of the waves.

Schizymenia Dubyi. Duby's Schizymenia.

Fronds from a few inches to a foot long, and three to five inches wide, commencing with a very short, flattened stem, which gradually expands into a thinnish, fleshy, membranous oval disc, waved at the edge. Spore-clusters very minute, enveloped in a transparent membrane, and disposed beneath the surface of the frond, slightly prominent; tetraspores not known.

This plant, formerly *Kallymenia Dubyi*, grows between the tide-marks in sheltered positions. It is annual, and in perfection in early summer. It is more rare than *Schizymenia edulis*, and is distinguished from that species by its shorter stem, and the much thinner substance of its frond. I recently received some very fine specimens of this handsome plant from that successful collector Mr. J. Gatcombe, who gathered them at Plymouth.

Genus XC. CATENELLA.

Frond nearly tubular, constricted, membranous, of a dull purple colour; the axis composed of a network of anastomosing, longitudinal filaments, from which are emitted forked, necklace-like, horizontal branches, whose tips, joined together with gelatine, form the outer wall of the frond. Spores in nuclei among the axile filaments of small, roundish, subsidiary branchlets.—CATENELLA, from the Latin *catenula*, a little chain.

A genus containing very few species of small insignificant plants, of which one only is a native of Great Britain.

Catenella opuntia. The cactus-like Catenella.

Fronds erect, from half an inch to an inch high, densely tufted, slightly branched, springing from a mass of creeping fibres; branches alternate or opposite, constricted, simple or forked, with pointed tips. Spores in round masses, contained in egg-shaped conceptacles with a terminal pore; tetraspores zonate, formed from the outer threads of the frond, and scattered among them.

Unlike most of its allies, this little plant is generally found on rocks which are not long submerged. It forms patches of two or more inches in diameter, and resembles slightly the young plants of *Chylocladia articulata*, but it may be readily distinguished by its darker colour and drier texture. It is perennial and not uncommon, but is rarely found in fruit.

Genus XCI. **GLOIOSIPHONIA.**

Frond very gelatinous, much branched, composed of an axis of interwoven jointed filaments, which form a longitudinal column at first, and subsequently a tube, and a periphery of numerous, whorled, necklace-like, forked filaments, set in gelatine. Spores numerous, enveloped in a gelatinous membrane, and immersed in clusters (favellæ) beneath the periphery; tetraspores not known.—GLOIOSIPHONIA, from the Greek *gloios*, viscid, and *siphon*, a tube.

The single species of this genus is widely distributed on the shores of the Atlantic Ocean, in both hemispheres. Its characters are very distinct, and the frond has a very curious structure, not easy to describe. Dr. Harvey speaks of it as wholly composed " of articulated confervoid filaments, invested with transparent gelatine ;" and with reference to the change which takes place in the axis, he remarks that it arises " either from distension or the perishing of the central cells."

Gloiosiphonia capillaris. The slender Gloiosiphonia.

Frond from a few inches to a foot long, filiform, much branched, very tender and gelatinous, composed of jointed filaments enveloped in transparent gelatine, when young comparatively solid, becoming tubular with age. Spores numerous, in clusters sunk in the branches beneath the outer series of filaments.

This is a very delicate and beautiful plant. It grows in deep water or in tide-pools. It is annual, and in perfection in summer. The recorded habitats are numerous all round our coast, but my own experience leads me to consider it a rare plant. It should be laid out in sea-water, or it will lose its colour.

Genus XCII. DUMONTIA.

Frond tubular, when young filled with a loose network of anastomosing filaments, which become obsolete as the plant matures and leave the tube empty; the wall of the tube is formed of an inner series of elongate, jointed cells, arranged lengthwise, with horizontal, forked, necklace-like branches, and an outer series consisting of a single row of small cells. Spores in roundish clusters (favellæ), formed out of and among the forked branches of the cells of the wall; tetraspores cruciate, also formed from the cells of the wall, but, of course, in separate plants.—DUMONTIA, in honour M. Dumont, a French naturalist.

The species of this genus are widely distributed, chiefly in temperate climates; only one is found on our coasts.

Dumontia filiformis. The thread-like Dumontia.

Fronds varying from an inch to two feet in length, and from one-tenth to half an inch in width; stem undivided; branches very long and simple, attenuate at the tip. Spores formed out of the forked cells of the wall of the frond.

This is a common species, and its characters are well marked and constant. Luxuriant specimens, with very wide fronds, whose branches are variously twisted and waved or frilled, are occasionally found near the mouths of fresh-water streams. This form is mentioned by Dr. Harvey as var. *β. crispata*. *D. filiformis* grows on rocks, etc., in tide-pools during summer, and is annual. Some specimens that I obtained in Jersey a year or two since were proliferous at the tips of the branches.

Order XIX. SPYRIDIACEÆ.

Red or Brown-red Sea-weeds with a thread-like, jointed, monosiphonous frond, more or less coated with small cellules. Fructification :—1. Spores formed in the upper cells of branched jointed threads, which radiate from a placenta enclosed in a cellular pericarp in external conceptacles; 2. Tetraspores external on the ramelli.

Genus XCIII. SPYRIDIA.

Frond thread-like, compressed, much branched, composed of a single, jointed, thick-walled tube, coated externally with small, many-sided cells, and beset with slender, minute, thread-like, jointed ramelli. Spores oblong, in clusters, several of which are enclosed in a membranous pericarp, in concep;acles borne at the ends of short branches; tetraspores ripartite, arranged singly along the branchlets.—SPYRIDIA, from the Greek *spuris*, a basket.

The six or seven species which are at present included in this genus are all natives of warm or temperate latitudes, and only one of them belongs to our marine flora.

Spyridia filamentosa. The filamentose Spyridia.

Fronds from a few inches to a foot long, and from once to twice as thick as a bristle, thread-like, much branched; branches alternate, repeatedly divided, thickly set with short, jointed, acute branchlets.

This species is widely distributed throughout the tropics of both hemispheres, and as far north as the south of England. Extending over so wide a range of latitude, it is not extraordinary to find it varying much in size and luxuriance; at the same time its general characters are distinct and sufficiently constant. It grows near low-water mark, is perennial, and in perfection in summer. It has been found in several localities on our south coast, and as far north as Anglesea. I have gathered it in moderate abundance in Jersey.

Order XX. CERAMIACEÆ.

Red or Brown-red Sea-weeds with a thread-like, jointed, more or less barked, one-siphoned frond; the bark, when present, formed of many-sided cells. Spores congregated in masses (favellæ) in transparent membranous sacs, which are either naked or surrounded by a collar of short branchlets; tetraspores external, formed either from the tips of the frond or the bark-cells.

Genus XCIV. MICROCLADIA.

Frond flattened, forkedly branched, composed of a jointed axis, not visible on the surface, surrounded by a thick bark, which is formed of two series of cells; those within large, angular; those without small. Spore-clusters (favellæ) seated on the branches, surrounded by a collar consisting of a few short branchlets.—MICROCLADIA, from the Greek *mikros*, small, and *klados*, a branch.

Microcladia glandulosa. The glandular Microcladia.

Fronds growing in tufts, from one to four inches high, somewhat flattened, much branched in an irregular manner; branches forked, the same width throughout, with very patent axils and short branchlets, which are either awl-shaped or bifid, and hooked at the tip. Spore-clusters somewhat globular, seated on the outer margin of the branches; tetraspores tripartite, or rarely cruciate, arranged in a line in the substance of the outer edge of the branchlets.

The first authentic specimen of this rare and interesting plant was found by Mrs. Griffiths on the coast of Devonshire in the year 1803. Other specimens have subsequently been found in the same neighbourhood, which appears to be the only locality in this country where it is to be obtained. My friend Mr. Stevens has collected it at Torquay, and Mr. Gatcombe at Plymouth, and I have to thank both these gentlemen for kindly supplying me with specimens. This species generally grows parasitically on other sea-weeds, at and beyond low-water mark. It is annual, and fruits in summer.

Genus XCV. CERAMIUM.

Frond thread-like, jointed, forked or pinnate, more or less coated with small roundish cells, which are irregularly arranged. Spore-clusters (favellæ) enveloped in a transparent sac, stalkless, set on the branches and surrounded by a collar of short branchlets; tetraspores tripartite, roundish, formed from and among the bark-cells, prominent on the surface of the frond.—CERAMIUM, from the Greek *keramos*, a pitcher.

A large number of species are included in this genus; but recent writers on the subject are not agreed as to whether certain forms are entitled to specific rank, or should be considered mere varieties. There are eleven British species which are pretty firmly established, and which represent the principal sections into which the genus is usually divided. These sections are:—

1. *Frond covered throughout with small bark-cells.*—This contains *C. rubrum*.
2. *Frond with nodes covered to a definite limit with bark-cells, the internodes being transparent.*—This contains *C. diaphanum; C. Deslongchampsii; C. tenuissimum; C. gracillimum; C. strictum;* and *C. fastigiatum.*
3. *Frond with nodes covered to a definite limit with bark-cells, and armed with one or more spines, the inter-*

nodes being transparent.—This contains *C. echionotum; C. acanthonotum;* and *C. ciliatum.*

4. Frond covered throughout with small bark-cells, and each node armed with a single spine.—This contains *C. flabelligerum.*

All the *Ceramia* possess very beautiful microscopic characters; and if these be carefully examined, most of the species may be identified without difficulty. The arrangement of the dark-coloured bark-cells round the nodes gives to certain species the appearance of being striped or variegated. The spines, or thorns, with which the fronds of some kinds are armed are peculiar,—so far, at least, as the British marine flora is concerned,—to plants of this genus. They are very minute, and can only be seen under the microscope; but their effect is very visible in the almost insurmountable difficulty that is experienced when an attempt is made to disentangle a bunch of fronds of a spinuliferous species. The arrangement of the fronds in this genus is very symmetrical, and so is that of the branches, which are fan-shaped, and seem to radiate from some common centre, terminating in bifurcate, hooked tips. There is scarcely any part of our coast where specimens cannot be obtained; and the variations in which some of the species indulge afford ample field for the exercise of all the observing faculties, in the selection of complete series of the different forms. It is, moreover, probable that there are even yet new species to be discovered or determined, and that patient observation may lead to the modification of the existing arrangement of the British species.

Ceramium rubrum. The red Ceramium.

Frond from two to ten inches long, varying in thickness

according to the size of the specimen, sometimes of greater diameter at the base than a hog's bristle, sometimes much less, always becoming gradually thinner towards the upper part, irregularly, forkedly branched; the ultimate branchlets incurved or hooked at the tips; the nodes of the stem and branches contracted; the internodes in the lower part of the frond about twice as long as broad; both nodes and internodes coated all over with coloured, cortical cellules. Spore-clusters on the sides of the branches, embraced by three or four short branchlets; tetraspores arranged round and sunk in the nodes.

This species is the most common of all the Red weeds found on our shores. It is annual, and more luxuriant in summer and autumn than at other seasons. It grows on rocks and stones, or parasitically on other weeds, and extends from near the extreme limit of high tide to some distance beyond low-water mark. Subject to so many vicissitudes, it is not remarkable that it should be liable to great variety of form; some of these are so distinct that they have been described and figured as separate species; but later research, and the comparison of extended series of specimens, have demonstrated that they are not entitled to specific rank. The following varieties are described by Professor Agardh, and I therefore insert them :—

Ceramium rubrum, var. decurrens.

Frond regularly forked, or furnished with forked branchlets; internodes with a narrow transparent stripe in the centre, caused by the absence of surface cellules.

This is one of the forms hitherto considered distinct. It is figured under the name of *C. decurrens* in Harvey's

'Phycologia,' but in his later works he has degraded it to a variety.

C. rubrum, var. proliferum.

Frond forked, beset on all sides with numerous simple or forked branchlets; nodes and internodes densely covered with cells. Spore-clusters generally destitute of branchlets.

This was formerly *C. botryocarpum*, figured plate 215, Harvey's 'Phycologia Britannica.'

C. rubrum, var. secundatum.

Similar to the last variety, but with secund branches.

C. rubrum, var. pedicellatum.

Fronds sparingly branched; branches scattered; nodes and internodes coated with coloured cellules.

Figured in Harvey's 'Phycologia,' plate 181.

Ceramium diaphanum. The transparent Ceramium.

Fronds from two to six inches long, as thick as a bristle below, becoming gradually thinner towards the upper part, irregularly forked; branches set with short, forked branchlets, with forcipate tips; nodes swollen, coated with purple cellules; internodes transparent, those of the main stem three or four times as long as broad. Spore-clusters near the tips of the branches, surrounded by involucral branchlets; tetraspores sunk in the nodes.

This is the largest and handsomest of our British *Ceramia*, and is very easily recognized by its numerous branches, its large size, and distinctly chequered stem. Its colour varies from a delicate pink to a dark purple.

It sometimes grows between tide-marks, but is generally parasitical on larger algæ, in deep water, whence specimens are washed on shore. When floating, it is so transparent as to be scarcely visible to any but a practised eye. Some of my finest specimens were found whilst bathing in Jersey, and I have vivid recollections of the difficulty of distinguishing them from the surrounding water.

Ceramium Deslongchampsii. Deslongchamps's Ceramium.

Frond two to four or five inches long, very slender throughout, forkedly branched; branches much divided, either naked, or set with simple or forked branchlets, which are sometimes alternate, sometimes secund, and occasionally crowded, making the upper part of the frond very bushy; tips of the branches straight and spreading; nodes coated with a band of coloured cellules; internodes transparent, those of the lower part of the stem about twice as long as broad, those of the branches shorter than broad. Spore-clusters sessile on the sides of the branches; tetraspores set round the nodes, large, and very prominent.

The long, slender branches, and straight tips of this species readily distinguish it from all other British *Ceramia*. It grows parasitically on various sea-weeds, and on the perpendicular sides of rocks, near low-water mark. It is of a deep-purple colour, and does not adhere firmly to paper unless steeped for a long time in fresh-water. The fructification of this plant is very anomalous. The apparent spore-clusters are not surrounded by involucral branchlets like those of all the other *Ceramia*, and, what is still more puzzling, they seem to grow on the same plant, sometimes on the same branch, as the tetraspores. They do not, moreover, contain spores of the usual form, but only a very fine powder.

Ceramium tenuissimum. The slender Ceramium.

Fronds tufted, two to five or six inches long, of the thickness of hair throughout, much and forkedly branched, with very wide axils; branches and branchlets spreading; tips slightly curled inwards; nodes swollen, coated with coloured cellules; internodes transparent, those of the middle of the stem from four to six times as long as broad, becoming shorter above. Spore-clusters surrounded by involucral branchlets, near the tips of short branches; tetraspores prominent, on the outer side of the nodes of short branches, one or more on each node.

There is a very close resemblance between this species and *Ceramium strictum*, but at the same time both possess sufficient distinctive characters to be identified with certainty. *C. tenuissimum* has its tetraspores arranged on the outer side only of the branchlets, one or two on each node, and the axils of the divisions of its fronds are very wide-spread. In *C. strictum*, on the other hand, the tetraspores extend all round the nodes, several in each, and the axils of the branches are comparatively acute.

Ceramium gracillimum. The very slender Ceramium.

Frond two or three inches long, excessively slender throughout, gelatinous and tender, irregularly divided in a forked or alternate manner; branches set with minute, forked, fan-shaped branchlets; tips incurved; nodes swollen, coated with coloured cellules; internodes transparent; those of the lower part of the stem many times as long as broad, those of the minute branchlets very much shorter. Spore-clusters in pairs, surrounded by long, forcipate, involucral branchlets; tetraspores projecting singly from the nodes.

In certain localities this species appears to be abundant, as, for instance, near Kilkee, on the coast of Ireland, where Dr. Harvey first discovered it, in 1844, and where it covered the rocks " almost to the exclusion of every other species, both in places left bare at low water and in the small tide-pools." Whether it be very local, or whether it be frequently overlooked on account of its small size, I do not know, but it is certainly considered a rarity by ordinary collectors. It grows parasitically on small algæ, is annual, and matures in autumn. It is so small, delicate, and gelatinous, that it is next to impossible to lay it out satisfactorily on paper.

Ceramium strictum. The straight Ceramium.

Fronds growing in dense tufts, from two to four inches high, very slender, and of nearly equal diameter throughout, irregularly, forkedly branched, having narrow, acute axils; branches and branchlets erect and straight; tips forcipate, and slightly hooked; nodes somewhat swollen, coated with minute, purple cellules, either quite smooth or clothed with long, transparent fibres; internodes three or four times as long as broad, transparent. Spore-clusters in the axils of the upper branches, embraced by a few short, involucral branchlets; tetraspores prominent, whorled round the nodes in the upper divisions of the branches.

The long, transparent fibres which clothe the nodes of

the upper branches of this species cannot, unfortunately, be relied on as a distinctive character, for in some specimens they are altogether absent; and, moreover, they are occasionally found on *C. rubrum* and other species. *C. strictum* must, therefore, be identified by the position of its tetraspores, its acute axils, and the general habit of the frond. It adheres closely to paper in drying, and its silky, silvery appearance renders it one of the most beautiful plants of the genus. It grows on shells, in tide-pools, near low-water mark.

Ceramium fastigiatum. The level-topped Ceramium.

Fronds from four to five inches high, very slender throughout, regularly forked, level at the top; axils acute; tips of the branches forcipate, slightly incurved; nodes coated with a definite band of small red cellules; internodes transparent, of a pale pink colour, those in the lower part of the frond very pale, and about six times as long as broad. Spore-clusters small, on the sides of the ultimate branches, supported by short, involucral branchlets; tetraspores on the outer edge of the branchlets, projecting from the nodes.

The dense, soft tufts of this exquisite little plant grow on rocks, etc., near low-water mark. They are annual, and in perfection late in autumn and winter. This latter fact, and their small size, which renders them likely to be overlooked, may account in part for the reputed rarity of the species.

Ceramium echionotum. The prickly Ceramium.

Frond three to six inches long, harsh and rigid to the

touch, very slender, of nearly equal diameter throughout, repeatedly forked; axils spreading, the branches frequently furnished with forked branchlets; tips strongly hooked inwards; nodes coated with a broad band of small, coloured cellules, and armed with irregularly-inserted, slender, colourless, one-jointed spines; internodes transparent, those of the lower part of the frond twice or three times longer than broad, those of the upper much shorter. Spore-clusters generally near the tips of the frond, or of the side branches, often axillary, surrounded by strongly incurved branchlets; tetraspores prominent on the outer edge of short branchlets, one or two in each node.

This is a common species, and very easily recognized under the microscope by the form and arrangement of the spines, which are always present on the nodes of the frond. It grows parasitically on small algæ, either in rock-pools or among pebbles, near high-water mark. It is annual, and in perfection in summer and autumn.

Ceramium acanthonotum. The one-spined Ceramium.

Fronds growing in dense, intricate tufts, slender, and of equal diameter throughout, repeatedly forked, fastigiate; the tips of the frond very strongly incurved; axils spreading; nodes coated with a broad band of minute, coloured cellules, armed on the outer edge with a single, robust, awl-shaped, coloured, three-jointed spine; internodes colourless, those in the lower part of the frond several times longer than broad, becoming very short at the upper part. Spore-clusters globose, on the sides of the branches, clasped by a single strongly-incurved, armed branchlet; tetraspores whorled round the nodes, very large and prominent, with a broad, transparent border.

The entangled tufts of this species are not uncommon, and possess comparatively easily recognized external characters; at the same time, all the species of this genus vary so much in different circumstances, and are so apt to resemble each other in outward appearance, that it is never advisable to trust the unaided eye to determine their identity. Every specimen should be submitted to the microscope, and the result of the examination recorded for future use.

Ceramium ciliatum. The ciliated Ceramium.

Fronds growing in dense tufts, very slender, and of nearly the same diameter throughout, rigid, repeatedly forked, with or without branchlets; tips of the frond very strongly incurved; axils spreading; nodes coated with a band of coloured cellules, and armed with a regular whorl of robust, awl-shaped, colourless, three-jointed spines; internodes transparent, those in the lower part of the frond several times longer than broad, becoming gradually very short in the upper part. Spore-clusters sessile on the sides of the branches, embraced by three or four involucral branchlets; tetraspores not very prominent, whorled round the nodes alternately with the spines.

The form of the spines, and the mode of their arrangement, are the distinctive characters of this species. The spines have three joints, the lower one much longer than either of the others. They are whorled round the nodes in a single, regular series, and all point towards the top of the frond. This plant grows in pools, or among stones, between the tide-marks, and is frequently parasitic. It is annual, and in season in summer.

Ceramium flabelligerum. The fan-bearing Ceramium.

Fronds growing in tufts, two to four inches high, tapering upwards, irregularly forked; branches fan-shaped, acute at the tip, forcipate, very slightly incurved, the outer edge of each node armed with a single, short, awl-shaped, three-jointed spine; the internodes of the lower part of the frond about twice the length of their diameter, those of the upper part about as long as broad; both nodes and internodes covered throughout with small, coloured cells. Spore-clusters attached to the upper branches, two or three together, supported by long, taper, involucral branchlets; tetraspores large, whorled round the internodes.

The cell-coated internodes, combined with the armed nodes, distinctly separate this species from all other *Ceramia*. *C. rubrum* possesses the former character; but no species has both. This is a rare plant, but I have collected it more than once in Jersey. It grows on sea-weeds, between high- and low-water mark, in summer and autumn.

Genus XCVI. PTILOTA.

Frond cartilaginous, compressed, much divided in a comb-like, pinnate manner, opake, composed of a single-tubed, jointed axis, surrounded by two layers of cells; those nearest the axis large, roundish, those on the surface minute, coloured. Spore-clusters nestling among involucral branchlets, at the tips of small branches; spores angular, numerous, enveloped by a transparent membrane; tetraspores tripartite, formed, either singly or several together, at the tips of the ultimate branchlets.—PTILOTA, from the Greek *ptilotos*, feathered.

The two species of this genus which are included in our flora are somewhat similar in external appearance and general character. They were formerly described as varieties of the same plant.

Ptilota plumosa. The feathery Ptilota.

Frond nearly flat, thicker in the middle than at the edges, very much branched; branches and branchlets pinnate, and opposite, of similar form, but of various sizes, a long and a short branch being frequently opposite to each other; the ultimate divisions of the frond comb-like, their teeth pointed, without apparent joints. Spore-clusters formed at the tips of the branchlets, nestling among several long, simple, acute, involucral branchlets; tetraspores at the tips of the teeth of the comb-like branchlets.

This is almost exclusively a northern species, and has not been found on our southern coasts. In the localities where it grows, the stems of the large specimens of *Laminaria digitata*, which are thrown on shore in summer and autumn, are frequently densely clothed with its dark-crimson, feathery fronds. When fresh it is a very handsome plant, and is an interesting object for the microscope. It shrinks much in drying, and does not adhere closely to paper, but, notwithstanding these disadvantages, it makes a beautiful specimen for a collection.

Ptilota elegans. The elegant Ptilota.

Frond three to six inches long, flaccid, very much branched, pinnately divided; all the younger pinnæ are jointed, and composed of a single row of large cells, obtuse at the tips; the larger branches are alternate, the lesser opposite, but not always the same length. Spore-clusters in pairs, naked, or surrounded by a few involucral branchlets; tetraspores on the tips of the ultimate branchlets, at first containing four sporules, subsequently eight or more.

In the 'Phycologia Britannica' this plant is called *P. sericea,* and Dr. Harvey gives very elaborate reasons for using that name, but in his more recent works he appears to acknowledge the prior claim of *elegans,* given by M. Bonnemaison, and adopted by Kützing and Agardh. I have, therefore, substituted *elegans* for *sericea,* although the latter name is that best known to English collectors. The fronds of this species are much narrower than those of *P. plumosa,* and its mode of growth is very different. It usually hangs in long tufts from the sides of perpendicular or overhanging rocks, and is not confined to any particular locality, but is generally distributed all round our coasts. It is perennial, and in perfection during summer and autumn.

Genus XCVII. DUDRESNAIA.

Frond cylindrical, very gelatinous, elastic, composed of three series of threads; the first or axial series loose netted, anastomosing; the second, closely packed, longitudinal, and the third or outer series horizontal, forked, necklace-like. Spores in globular masses, attached to the bark threads; tetraspores zonate, external, at the ends of the branches.— DUDRESNAIA, in honour of M. Dudresnay.

Under the new arrangement, which removed *D. divaricata* into a different Order, this genus contains only one British species, confined to southern localities.

Dudresnaia coccinea. The red Dudresnaia.

Frond bright rosy-pink, from six to ten inches long, tender, extremely gelatinous, much and irregularly branched;

branches alternate, bearing a second and third series of branchlets. Favellidia large, lodged at the base of the peripheric fibres of the frond near the end of the branches; tetraspores at the tips of similar fibres.

In a young state the branches of this plant are distinctly visible to the naked eye, but when mature the whole frond appears to be composed of gelatine, and when taken from the water becomes an undistinguishable mass of pink mucilage, with scarcely any indication of branches, and still less of internal structure. It requires peculiar treatment in drying, and should not be soaked in fresh water, nor subjected to any pressure until the second or third day after it is laid on the paper. It does not present a very promising appearance during the first stage of this process, but the collector must not be disheartened, for the result will probably prove more satisfactory than he anticipates. This is a deep-water species, but from its great weight is very liable to be dislodged by the sea, and I have found it in considerable abundance in Jersey, either floating, or thrown up on the sands. It is a summer annual.

Genus XCVIII. CROUANIA.

Frond gelatinous, single-tubed, jointed, the joints of the stem and branches whorled with numerous, minute, forked, jointed, level-topped branchlets; "favellidia subsolitary, near the apex of the ramuli, affixed to the base of the whorled ramelli, and covered by them, containing within a hyaline, membranaceous perispore, a subglobular mass of minute spores;" tetraspores cruciate, large, at the base of the branchlets.—CROUANIA, in honour of the brothers Crouan, of Brest, celebrated among French algologists.

These small gelatinous sea-weeds resemble the fresh-water *Batrachospermæ* more nearly than any salt-water genus. One species only has been found on our coasts, and that one very rarely.

Crouania attenuata. The attenuated Crouania.

Frond tufted, one to two inches long, very delicate and gelatinous, much branched; branches attenuate, jointed, the whole frond whorled at the nodes with minute, forked, fastigiate branchlets. Tetraspores cruciate, solitary; favellæ not seen on British specimens.

This exquisite little plant is extremely rare, and has never been collected in great abundance anywhere. It is parasitic on small sea-weeds, and has been found on *Cladostephus spongiosus*. When young the ramelli appear to clothe the stem and branches like a bark, but as the joints lengthen they separate into distinct whorls. The tetraspores are very large in proportion to the size of the plant.

Genus XCIX. HALURUS.

Frond thread-like, jointed, single-tubed, irregularly divided, with short, incurved, forked branchlets whorled round all the nodes. Spore-clusters borne on the tips of shortened branches, surrounded by involucral branchlets; spores angular, numerous, contained in a transparent envelope; tetraspores spherical, tripartite, borne on the inner sides of the forked branchlets of an involucre.

The plants which compose this genus were formerly included with the *Griffithsiæ*. Only two species have been described, and but one of these is found on our shores.

Halurus equisetifolius. The Equisetum-leaved Halurus.

Frond from four to eight inches long, irregularly branched; the branches alternate, set with one or more series of side branchlets; stem and branches thickly clothed with short, forked, jointed, incurved ramelli, which on the younger parts of the frond are regularly whorled round the nodes. Spore-clusters at the tips of shortened branches; tetraspores inside the involucral branchlets.

When young and freshly gathered, this is a handsome plant, but it loses its colour, and becomes coarse and shaggy with age. With the exception of the difference of colour, it bears a strong resemblance to *Cladostephus spongiosus*. It grows abundantly on our southern shores, but is much less luxuriant towards the north. It adheres firmly to paper, but, like all the species of *Griffithsia*, it must be laid out in salt water, or it will stain the paper.

Halurus equisetifolius, var. simplicifilum.

"Stems slender, irregularly branched, whorled with imbricated, straight, once-forked ramelli."

In the 'Phycologia Britannica,' this form is hesitatingly described as a distinct species. Agardh treats it only as a variety, and I have done the same, as I have never been able to find any distinct specific characters in the specimens I have examined.

Genus C. GRIFFITHSIA.

Frond single-tubed, jointed, forked. Spore-clusters (*favellæ*) surrounded by numerous, regular, involucral branchlets, containing many angular spores enclosed in a gelatinous sac (*periderm*); tetraspores spherical, at length tripartite, attached to the inside of involucral branchlets. —GRIFFITHSIA, in honour of Mrs. Griffiths.

This is a large and interesting genus, which is handsomely represented on our shores. The species are all of delicate structure, and must always be laid out in salt water, as they rapidly decompose in fresh.

Griffithsia setacea. The bristly Griffithsia.

Fronds growing in tufts, from three to eight inches or more long, jointed, forked, tapering gradually from the base to the tips; articulations many times longer than broad below, becoming gradually shorter above. Spore-clusters (*favellæ*) oval, enveloped in a transparent membrane, arranged among short-stalked, globular tufts of forked, involucral branchlets; tetraspores attached to the inner side of involucral branchlets, which form a globular tuft, similar to that containing the favellæ.

This is the most common species of the genus, and grows on all our coasts in more or less abundance. It is of a bright red colour, and very crisp when fresh, but it becomes flaccid on exposure to the air, and when put into fresh water its cells discharge the crimson colouring-matter with which they are filled, and the whole plant assumes the dull orange hue indicative of decay. Dr. Harvey ascribes to this plant a quality which will prove valuable to the possessors of marine aquaria. He writes:—" Delicate as the structure of this plant assuredly is, no marine alga is more patient of confinement, or may be more easily domesticated. A tuft placed in a closed bottle of sea-water in April, 1846, is now, after more than two years' imprisonment, apparently as fresh and healthy as when first taken from the sea. The water has not been changed, and is perfectly clear and pure. The plant has not grown much, as the bottle is a small one, but its threads reach nearly to the surface of the water, and no decay has taken place."

Griffithsia secundiflora. The side-fruited Griffithsia.

Fronds growing in tufts, from four to eight inches long, with a fan-shaped outline, rather gelatinous but firm, jointed, forked, blunt at the tips, with short, horizontal ramuli occasionally issuing from the lower branches; articulations two to four times as long as broad, with a wide, transparent margin. Favellæ not known; tetraspores attached to the inner sides of forked, incurved, involucral branchlets, which are arranged in the form of an umbel at the tips of very short side branches.

This species slightly resembles *G. setacea* in appearance, but it may easily be distinguished by its larger size, and the blunt tips of its branches. I can find no record of fruit on any British specimens, and though I have gathered it somewhat abundantly in Jersey in the months of June, July, and September, all my specimens are perfectly barren. I may add, that those collected in September are thicker, and of a richer colour, though not so long as those gathered earlier in the year. I have generally found them growing erect, at the bottom of shallow, sandy pools. They adhere firmly to paper, but, like all other Griffithsias, must be laid out in salt water.

Griffithsia corallina. The coral-like Griffithsia.

Fronds tufted, from two to six or eight inches long, very

gelatinous, repeatedly and regularly forked; articulations two to four times longer than broad, tapering from the base to the apex, which is rounded, those below somewhat cylindrical, those in the middle almost pear-shaped, those above connected in necklace-like strings. Favellæ stalkless, surrounded by short branchlets at the apex of the articulations; tetraspores in clusters, several of which are whorled round the joints, and surrounded by short, involucral ramelli.

The appearance of this species differs from that of all other British sea-weeds. It grows on rocks in deep pools, near low-water mark, and is distributed all round our coasts. It is annual, and in perfection in summer. Its natural colour is a rosy-crimson, but this is rapidly lost by exposure, or in fresh water; indeed it is very difficult to obtain specimens perfect in this respect, as the collector is for the most part compelled to be content with those which may be thrown on shore by the waves. Even faded specimens, however, are very beautiful when fresh, their transparent, gelatinous texture causing them to sparkle in the sun, like clusters of brilliant beads; and it is with a feeling of regret that one applies the pressure that will reduce them from this lovely phase of their existence to the mere shadow of their former selves, which will appear on the paper when the process of drying is completed.

Griffithsia Devoniensis. The Devonshire Griffithsia.

Fronds tufted, from two to three inches long, very slender, forkedly divided, gelatinous; articulations many times longer than broad, cylindrical, slightly thickened and constricted at the joints. Favellæ not known; tetraspores on the inner face, of short, involucral branchlets, which are densely whorled round the joints of the main stem.

This species is very rare, and confined to southern localities, where it grows on mud in deep water, and must therefore be obtained by dredging, or when floating or cast up by the waves. It resembles *G. barbata*, but is separated from that species by the position of the tetraspores, and by the less uniform length of the branches.

Griffithsia barbata. The bearded Griffithsia.

Fronds tufted, very slender, repeatedly and regularly forked; articulations from five to eight times as long as broad, slender at the base, becoming thicker and rounded at the apex, those of the terminal branches giving off slender, byssoid, forked, spreading branchlets. Favellæ in pairs, stalked, formed out of truncate branches, and surrounded by numerous, simple or forked, involucral branchlets; tetraspores spherical, attached singly to the branchlets, which spring from the upper branches.

Though far from common, this species is less rare than the last. It grows parasitically, is annual, and fruits in summer. I have found it in great abundance in Jersey, floating at the edge of the tide. I have also seen specimens from Brighton and other localities. When laid out on paper, the frond is fan-shaped, and the tuft forms nearly a perfect circle.

Genus CI. **SEIROSPORA.**

Fronds rosy, thread-like; stem and branches jointed, one-tubed, veined. Fructification, tetraspores disposed in necklace-like strings, at the tips of the branches.—SEIROSPORA, from the Greek *seira*, a chain, and *sporos*, a seed.

This genus was founded by Dr. Harvey, for the reception of a new British species, discovered by Mrs. Griffiths in 1838. He thus alludes to it in the 'Phycologia Britannica':—" I was not so confident of its claims to this distinction [that of being a new species], and first described it as a variety of *Callithamnion versicolor*, chiefly remarkable for a curious modification of fruit. There is, indeed, a close resemblance to strong-growing plants of *C. versicolor*, so close that we are driven to look to the fructification for marks of difference. Here, however, the characters are so broadly defined, that if we regard the fruit of our *Seirospora* as being normal, according to the view first taken by Mrs. Griffiths, and latterly, though with some hesitation and reluctance, adopted by me, we shall be compelled to form a new genus for its reception. In *Callithamnion*, the tetraspores are borne laterally along the ramuli; here the ramuli themselves are converted at maturity into strings of tetraspores,—a tetraspore being formed within each of the articulations of the ramulus. This character is quite as strong, in a generic view, as that which separates any other genus of *Ceramiaceæ*, and amply sufficient to distinguish the plant from *Callithamnion*." Professor Agardh, on the other hand, maintains that the plant is a true *Callithamnion*, and describes it as *C. seirospermum*.

Seirospora Griffithsiana. Mrs. Griffiths's Seirospora.

Fronds growing singly or in tufts, from one to three inches

high; stem bristle-like, jointed, opake, veined; branches numerous, long, simple, alternate, spreading, long below, becoming shorter towards the apex of the plant, often bearing a second series; plumules much divided, jointed, with a narrow, egg-shaped outline. Tetraspores tripartite, elliptical, formed in necklace-like strings from the upper branchlets of the plumules.

This beautiful plant was first discovered on the coast of Devonshire, and has since been collected in Scotland, Ireland, the Channel Islands, and likewise in Sweden. It grows in deep water, and is washed on shore in summer. Its substance is flaccid and gelatinous, and it adheres closely to paper.

Genus CII. **CORYNOSPORA.**

Frond filiform, dichotomous, beset with pinnate branchlets, contracted at the joints, one-tubed. Favellæ near the tips of the side branches, girt with curved, involucral branchlets, containing many angular spores; tetraspores solitary, on short stalks, on the axils of the branchlets.—CORYNOSPORA, from the Greek *koryne*, a club-like shoot, and *sporos*, a seed.

The only British species belonging to this genus was formerly included in *Callithamnion*; the separation is founded on the mode of fructification. There are four or five Continental species.

Corynospora pedicellata. The pedicellate Corynospora.

Stem bristle-like, transparent, irregularly branched; branches long, either simple or repeatedly divided in a somewhat forked manner; branchlets alternate, twice or thrice dichotomous. Favellæ unknown; tetraspores borne on short stalks in the axils of the branchlets.

This species is not uncommon on the south and west shores of England and Ireland, but is more rare further north. It grows on rocks and submerged woodwork, or in deep water, and is annual. It adheres closely to paper, and is very beautiful when dry ; but specimens gathered in autumn are frequently covered with Diatomaceæ, and have not the bright colour which characterizes the plant in summer.

Genus CIII. CALLITHAMNION.

Fronds thread-like, branched, jointed, one-tubed; the stem and branches in some species made opake by the development of decurrent fibres in the walls of the primary cells. Spores angular, contained in a transparent envelope, in favellæ, which are generally in pairs, stalked or axillary on the branches; tetraspores tripartite or cruciate, oblong or globose, naked, sessile or stalked, distributed among the branches.—CALLITHAMNION, from the Greek *kalos*, beautiful, and *thamnion*, a little shrub.

More than a hundred species are included in this genus, and about a quarter of the number are natives of our shores. They are, with one or two exceptions, of small size, and some of them so tiny as to be scarcely visible to the naked eye. The genus is variously subdivided by different writers, some of whom go so far as to arrange its constituents in three separate genera. I have adopted the principal sections used by Professor Agardh, which are at the same time simple and sufficient for the purposes of a popular work. Nearly all the species are distinguished by microscopic characters, and several of them rest on very slight and uncertain foundations. They are, moreover, very variable, and are altogether more difficult than those of perhaps any other genus to determine with certainty. The rosy colour, feather-like form, and delicate texture of the fronds of most of the species cause them to be conspicuous for beauty, even in an assemblage which includes such potent rivals as are to be found among the Red series of British sea-weeds.

SECTION 1.—*Frond shrub-like; the stem and branches veined, indistinctly jointed; branches pinnate; branchlets alternate, or on one side only of the branch.*

Callithamnion arbuscula. The bush Callithamnion.

Fronds growing several from the same base, from two to six inches high; stems naked below, without visible joints; branches alternate, those of the second series thickly set with minute plumules, which are densely crowded towards their tips; plumules furnished with alternate, simple or forked, spreading, recurved pinnules. Favellæ in pairs, springing from the stem of the plumules; tetraspores spherical, stalkless, plentifully produced on the upper edge of the pinnules of the plumules.

This is one of the most robust of British Callithamnia, and thrives in the roughest water, and on the most exposed rocks. It grows abundantly on the west coasts of Scotland and Ireland, and is rare in England. It is perennial, and fruits in summer and autumn.

Callithamnion spongiosum. The spongy Callithamnion.

Fronds from two to four inches long, flaccid, spongy; stem with indistinct joints, veined, much branched; branches of two or three series, long, thickly clothed with quadrifarious, round-topped plumules, which are furnished with alternate, several times forked pinnæ, with short, bifid, very obtuse tips. Favellæ large, generally in pairs, near the tips of the plumules; tetraspores tripartite, solitary, stalkless, in the axils of the pinnæ of the plumules.

Although the colour of this species is never very brilliant, and is easily deteriorated by exposure, the form of the plant itself is so beautiful when perfect that it always makes a handsome specimen. It must be laid out immediately it is gathered, for it decays very rapidly. It grows near low-water mark, either on rocks or sea-weeds, is annual, and fruits in summer.

Callithamnion Brodiæi. Brodie's Callithamnion.

Fronds growing in tufts about two inches high; stem slender, opake below, more or less jointed above, veined; branches alternate, spreading, veined, those below longest, becoming gradually shorter upwards, the second or third series furnished with alternate plumules; plumules pinnate, with alternate, simple, spreading pinnules, from the inner side of the tips of which spring a few awl-shaped processes. Favellæ large, in pairs on the sides of the lesser branches, which are frequently distorted; tetraspores oval, stalkless, on the processes of the pinnules.

This is peculiarly a British species, and has not been found in any other country. It is parasitic, and grows near low-water mark, is annual, and fruits in summer.

Callithamnion tetragonum. The four-angled Callithamnion.

Fronds from two to five inches long; stem indistinctly jointed, veined; branches simple or alternately divided, clothed with short, alternate, spreading, level-topped plumules; branchlets of the plumules incurved, robust, tapering suddenly at the point. Favellæ in pairs, formed in the centre of the plumules; tetraspores tripartite, very minute, on the inner side of the tips of the branchlets of the plumules.

This is a handsome plant, and, for the genus, of large size; it is parasitic on various Algæ, and, in a young state, forms a fringe on the edge of their fronds; when mature, the form of the plant is pyramidal. In drying, the specimens lose much of their beauty: the plumules become matted together, and do not lay out well, nor adhere to paper.

Callithamnion tetragonum, var. β. brachiatum.

Branchlets of the plumules more slender, and tapering from the base to the tip.

This variety is figured and described as a separate species (*Callithamnion brachiatum*) in the 'Phycologia Britannica,' but it would appear that Dr. Harvey adopted this course in deference to Professor Agardh, and as the latter botanist has since altered his opinion, I feel myself at liberty to insert the plant in what I believe to be its correct position, as a variety.

Callithamnion tetricum. The rough Callithamnion.

Fronds two to eight inches long, very bushy, much

branched; branches robust, either simple or divided, alternate, densely clothed with shaggy, coarse, irregularly divided branchlets, those of the upper divisions of the frond comparatively long and slender. All the branches are crowded with narrow pinnate plumules, furnished with spreading acute pinnules. Favellæ in pairs on the pinnules of the plumules; tetraspores tripartite, minute, stalkless, on the inner sides of short processes which issue from the pinnules of the plumules.

Were it not for a difference in the colour, this species might be more readily mistaken for *Sphacelaria scoparia* than for any species of *Callithamnion*, and it is difficult to believe that there is any affinity between its coarse shaggy fronds, and those of its more delicate allies. It is common, and grows on the perpendicular sides of rocks left bare by the receding tide, from which it hangs in long taper tufts of a dull red colour.

Callithamnion Hookeri. Hooker's Callithamnion.

Fronds growing singly or several together, from one to four inches long; stem indistinctly jointed, veined; branches alternate, spreading, flexuous, set with a second and third series of branchlets, which are naked below and clothed with nearly horizontal spreading plumules above; the joints throughout the plant are twice as long as broad. Favellæ in pairs, without stalks on the plumules; tetraspores spherical, numerous, on the joints of the inner edge of the pinnæ of the plumules.

This species is named after the late Sir William Jackson Hooker, of Kew. It is pretty generally distributed all round our coasts from north to south; it grows parasitically on other Algæ, and on rocks near low-water mark, and is annual.

Callithamnion fasciculatum. The fasciculate Callithamnion.

Fronds tufted; branches erect, level-topped; plumules elongate, erect, linear-obovate, truncate; pinnæ long and flexuous, the lowermost simple, appressed, the upper branched, spreading; articulations veined, from twice to four times as long as broad, those of the pinnæ contracted at the joints. Favellæ on the joints of the upper branches; tetraspores tripartite, solitary, near the base of the pinnules.

Only a single specimen of this species is known to have been found, and it is therefore impossible to decide with certainty that it is distinct. It is figured by Dr. Harvey in the 'Phycologia Britannica,' and Professor Agardh inserts it in his ' Species Algarum ;' I have, therefore, copied the description given in the former work.

Callithamnion Borreri. Borrer's Callithamnion.

Fronds growing in tufts, from one to four or five inches long, much branched from the root in a fan-shaped manner, distinctly jointed throughout; branches set with two or three series of spirally inserted lesser branches, naked below, plumulate above; plumules bare for about half their length, set with simple pinnæ for the remainder, the lower pinnæ longest, becoming gradually shorter towards the apex; articulations of the branches two to five times, of the branchlets about twice as long as broad; throughout the frond the various divisions spring from each joint, and are regularly alternate. Favellæ in pairs, without stalks, on the plumules; tetraspores roundish, on the inner side of the pinnæ of the plumules, one or several on each; some of them contain eight grains, each of which, when ripe, becomes a tripartite tetraspore.

This is a handsome, feathery plant, of a bright carmine colour; it grows near low-water mark, is annual, fruits in summer, and is rather rare. It may be readily distinguished by the long bare stem and simple pinnæ of its plumules. It is named in honour of Mr. Borrer, a well-known botanist, by whom it was discovered.

Callithamnion polyspermum. The many-spored Callithamnion.

Fronds forming globular tufts one to three inches in diameter, slender, delicate, much branched; branches zigzag, of two or three series, plumulate from each joint; plumules alternate, pinnate, or rarely bipinnate; pinnæ alternate, short, simple, spine-like; articulations of the branches four or five times, of the branchlets about twice as long as broad. Favellæ of large size, in clusters, on the rachis of the plumules; tetraspores tripartite, borne on the inside of the pinnæ of the plumules; antheridia formed of innumerable minute cells, strung together, occupying the same position as the tetraspores.

Although extremely delicate and slender, this plant sometimes grows so abundantly on *Fucus serratus* and *Fucus vesiculosus* as to completely cover their fronds. It is one of the most common of the *Callithamnia*, and may be found in summer in almost every locality round our coasts. The short, awl-shaped pinnules of the plumules are its most obvious character; but it is generally a variable plant, and must be carefully examined under a microscope, in order to determine its identity.

Callithamnion tripinnatum. The thrice-pinnate Callithamnion.

Fronds growing in tufts, from one to two inches high, membranaceous, fan-shaped, about thrice pinnate; plumules pinnate below, bipinnate above; the pinnæ short and simple, becoming gradually longer and branched to the middle of the plumule, and thence gradually shorter to the tip; pinnæ and pinnules alternate, the former naked for the first half of their length, with the exception of the joint next the stem, which usually bears a short pinnule; joints of the stem and branches about thrice as long as broad. Favellæ unknown; tetraspores tripartite, on the upper sides of the basal and ultimate pinnules.

This exquisite little plant is unfortunately very rare. It grows on rocks at extreme low-water mark, is annual, and in perfection at the end of spring. Its distinctive character is the solitary, generally fertile pinnule, that is borne on the first joint of each pinna of the plumule.

Callithamnion affine. The allied Callithamnion.

Fronds growing in tufts, from two to three inches high, feather-like, much branched; stem opake, veined; branches long, with a roundish outline; plumules alternate, short, simple, awl-shaped; articulations of the branches three or four times, of the pinnæ about once as long as broad. Favellæ in pairs, near the tip of the plumules; tetraspores tripartite, solitary, near the base of the pinnules.

The title of this plant to specific rank is of the most doubtful nature. It partakes of the characters of the normal states of three or four species, and is scarcely separable, with any degree of certainty, from some of their intermediate forms. Both Harvey and Agardh have hitherto retained it, and therefore I insert it here.

Callithamnion thuyoideum. The cypress Callithamnion.

Fronds growing in tufts, from one to three inches long; branches alternate, spreading, those at the base of the frond longest, the rest becoming gradually shorter; plumules alternate, triply pinnate, borne on each articulation of the branch, the first being always on the upper side of the rachis; articulations of the stem and branches variable, of the plumules uniformly about twice as long as broad. Favellæ solitary or in pairs, on the stems of the plumules; tetraspores tripartite, minute, on the tips of the ultimate series of pinnules.

This is another very beautiful and too rare species. It grows on rocks near low-water mark in spring and summer, and is annual. Several localities on the south coast, Yarmouth, Swansea, and the western shores of Ireland, are its recorded habitats. Its colour is bright pink, and its mode of growth very compact.

Callithamnion gracillimum. The very graceful Callithamnion.

Fronds growing in tufts, one to four inches high, very slender, irregularly branched; main branches few, simple, with an ovate outline, attenuated at the point; the lower plumules short, pinnate, the upper long, lanceolate, thrice pinnate; all the divisions alternate. Favellæ roundish, or irregularly lobed, near the base of the plumules; tetraspores minute, on the tips of shortened pinnules of the plumules.

This very elegant species is extremely rare; it was first found in this country by Mrs. Griffiths, growing on the mud which covers the base of the pier at Torquay, and was subsequently gathered at Milford Haven and Falmouth. It is annual, and attains perfection in

summer. There is a close resemblance between it and
C. thuyoideum, they both bear their tetraspores at the
tips of the pinnules, and in this character differ from
all the allied species.

Callithamnion corymbosum. The corymbose Callithamnion.

Fronds solitary, or growing in tufts, distinctly jointed
throughout, much branched; stem and branches as thick as
a bristle below, extremely slender above; plumules very
fine, flaccid; pinnæ opposite, alternate or secund (on one
side only), the lesser and ultimate divisions being forkedly
divided, and level-topped; articulations of the branches
sometimes veined, six or eight times as long as broad, of the
branchlets shorter. Favellæ in pairs, on the rachis of the
plumules; tetraspores tripartite, solitary, near the axils of
the forked pinnæ of the plumules.

This species is very variable, and was formerly divided
into two: the short-jointed, pinnately-branched, robust
form being called *C. versicolor*. Dr. Harvey pointed
out the error of this arrangement; and any one who will
take the trouble to examine an extended series of specimens will be convinced, I think, that the two forms
belong to the same species. *C. corymbosum* grows on
Zostera, sea-weeds, or rocks, near low-water mark. It
is annual, and in perfection in summer.

Callithamnion byssoideum. The byssus-like Callithamnion.

Fronds growing in dense tufts, from one to three inches
high, extremely fine, flaccid, tender, jointed nearly to the
base, much branched; lower branches irregularly divided,

upper plumulate; plumules slender, distantly once or twice pinnate; pinnæ alternate, or on one side only, slightly tapering, often branched at the tip; articulations of the branches six or eight times as long as broad, of the branchlets three to six times. Favellæ in pairs, generally terminating truncated branches; tetraspores tripartite, large, stalkless, on the pinnules of the plumules.

The substance of this species is very gelatinous and tender, and the natural colour of perfect specimens a brilliant pink. It grows on Algæ in tidepools, especially on *Codium tomentosum*. It is annual, and of frequent occurrence. In external appearance it resembles *C. corymbosum*, but its mode of branching, and other characters, are sufficiently distinct when examined under a microscope.

Callithamnion interruptum. The interrupted Callithamnion.

Fronds growing parasitically on Algæ, much branched; branches alternate, somewhat lanceolate. Tetraspores cruciate, on short stalks, on the inner side of the axils of the branches.

This species is figured in the 'English Botany' as *Conferva interrupta*, and is retained by Professor Agardh in his 'Species Algarum,' but is omitted from the 'Phycologia Britannica.' I do not possess a specimen, nor have I been able to borrow one. Nevertheless, I do not like to omit a plant that has been recorded as British, and whose identity with any other species I cannot satisfactorily establish.

Callithamnion roseum. The rosy Callithamnion.

Fronds growing in dense tufts, three or four inches long;

stems in young plants transparent, becoming opake and veined as they increase in age, much branched; branches alternate, much divided; plumules alternate, simply pinnate, with a roundish or ovate outline; pinnæ long, more or less incurved, either quite simple, or furnished with a few pinnules at the tip; articulations of the stem and branches four or five times as long as broad, somewhat swollen at the joints, those of the plumules becoming gradually shorter. Favellæ in clusters, generally terminating short branches; tetraspores tripartite, borne on the upper side of the pinnules of the plumules, one at the apex of each of the three or four lower joints.

This is a very handsome, dark-coloured species, not uncommon in many localities. It grows on rocks and large sea-weeds, near low-water mark, or at the mouths of tidal rivers. It is annual, and matures in summer.

SECTION 2.—*Fronds pinnate, the pinnæ opposite.*

Callithamnion floccosum. Pollexfen's Callithamnion.

Fronds densely tufted, from one to four inches in length, flaccid, irregularly divided in a distantly alternate, forked manner; branches naked, or clothed at intervals with short, secondary branches; articulations throughout the frond transparent, from twice to four times as long as broad, each bearing near its apex a pair of short, awl-shaped, spine-like branchlets. Favellæ unknown; tetraspores tripartite, stalked, near the base of the awl-shaped branchlets.

This is a very rare plant, apparently exclusively confined to northern latitudes. The Orkney Islands and Aberdeen are the recorded habitats in this country. It grows on rocks, near low-water mark, in spring, and is annual. The long, slender, comparatively naked branches, and the short, awl-shaped branchlets are characters which will readily reveal its identity to the collector who may have the good fortune to discover it.

Callithamnion plumula. The little feather Callithamnion.

Fronds growing in tufts, from two to six inches in length, flaccid, distichously branched; branches alternate or forked, repeatedly divided; plumules once or twice pectinate on their upper side, springing in pairs, or, in luxuriant specimens, in threes and fours, from near the apex of every joint of the stem and branches, about half a line to a line in length. Favellæ in clusters, on the tips of the main branches, which are always shortened; tetraspores cruciate, on the tips of the comb-like pinnules of the plumules.

This is one of the most common, easily recognised, and elegant species of this beautiful genus. The comb-like plumules clothing every joint of the plant are so visible to the naked eye, that a microscope is scarcely necessary to determine its identity. It grows on rocks and sea-weeds near low-water mark, and also in deep water, whence it is often washed on shore. Specimens differ occasionally in the luxuriance of the plumules; in some they are short and thick, in others long and very delicate. The colour varies from carmine to light brown; the substance is soft and tender, and the plant adheres closely to paper when dry.

Callithamnion cruciatum. The crossed Callithamnion.

Fronds growing in tufts, one to two inches long, flaccid,

sparingly and irregularly branched; branches irregularly divided, jointed, plumulate from every joint; plumules from half a line to a line long, opposite, or in threes and fours, very crowded at the tips of the branches; pinnules opposite, slender, cylindrical, blunt at the tips. Favellæ not known; tetraspores cruciate, elliptical, borne on the lowest joints of the plumules, either sessile or on very short stalks.

This plant occurs in many localities, but only in small quantities. The coasts of Devon, Wales, and Ireland are the chief recorded habitats. It grows on mud-covered rocks near low-water mark, and is annual.

Callithamnion pluma. The feather Callithamnion.

Fronds feather-like, rising from creeping filaments, from a quarter of an inch to about half an inch high, simple or alternately branched; plumules naked below, clothed above with short, opposite pinnules issuing from each joint; articulations of the stem about three times as long as broad, of the plumules about as long as broad. Tetraspores globose, tripartite, either borne on short, special stems near the base of the pinnules, or at the tip of a shortened pinnule.

This little plant is rare, probably because it is frequently overlooked. It grows parasitically on the stems of *Laminaria digitata,* on which its upright fronds are set so closely together that the patches which they form resemble crimson velvet. Some of the plumules have pinnules only on one side of the stem. It is annual, and in perfection in summer.

Callithamnion Turneri. Turner's Callithamnion.

Fronds rising at right angles from creeping fibres, which

attach themselves by small disc roots to the sea-weed on which the plant is parasitic, growing in tufts one to two inches high, simple, or repeatedly branched; branches opposite, sometimes alternate, spreading; branchlets simple, slender, opposite; articulations of the stem five to ten times as long as broad, of the branches four to five. Favellæ two-lobed, borne on shortened branches, and surrounded by involucral branchlets; tetraspores tripartite, globose, on the sides of short, simple, or branched stalks, which rise from the base of the branchlets.

This species is common all round our coasts. It grows on Algæ in tide-pools in summer, and is annual. Its habit and texture are very different from the allied species, and it may, therefore, be readily recognized.

Callithamnion barbatum. The bearded Callithamnion.

Fronds forming dense intricate tufts, matted together, one to two inches high, much and irregularly branched; branches alternate or opposite, long, simple, or bearing another series, clothed with minute, spine-like, opposite branchlets, which are distant or absent in the lower part of the branch, but closely and regularly placed near the tip; articulations twice or thrice as long as broad. Favellæ unknown; tetraspores tripartite, elliptic, with a broad transparent margin, borne on the branchlets.

Mr. Ralfs first gathered this species in 1838, and only one or two specimens have been found since; I fear, therefore, it must be considered very rare. It is supposed to be perennial, and grows on mud-covered rocks in tide-pools.

SECTION 3.—*Fronds without a distinct stem; branches scattered.*

Callithamnion floridulum. The little-florid Callithamnion.

Fronds forming very dense, nearly hemispherical tufts, very silky and slender, much branched in a forked or irregular manner, level-topped; branches few, long, erect, straight, those in the lower part of the frond long, becoming shorter above; branchlets few or none, closely pressed to the branches; articulations three times as long as broad. Favellæ not known; tetraspores tripartite, oval, on short stalks, arranged on one side of the branches.

On the west coast of Ireland large surfaces of rock are covered with this plant, and it is washed on shore in such abundance that the country-people use it for manure. It occurs in England, but is far from common.

Callithamnion Rothii. Roth's Callithamnion.

Fronds growing in dense velvet-like patches on the surface of rocks, about half an inch high, very slender, level-topped; branches simple, erect, longest below, nearly bare of branchlets; articulations twice as long as broad. Tetraspores tripartite, two, three, or four together on the end of a short branchlet near the tips of the branches.

This very small and slender species grows on rocks midway between high and low-water marks, or more rarely almost beyond the reach of the tide. It is perennial, and flourishes in winter and in cold climates. The arrangement of the tetraspores is its distinctive character.

Callithamnion mesocarpum. The middle-fruited Callithamnion.

Fronds from an eighth to a quarter of an inch high, rising from creeping filaments, erect, simple, or sparingly branched; branches alternate, naked, or with a few branchlets; articulations several times longer than broad, with a transparent margin. Favellæ unknown; tetraspores tripartite, elliptical, borne on the side branches.

Captain Carmichael discovered this minute plant at Appin, and described it as "growing on rocks in continuous tufts, forming a broad, shaggy, purple crust." The original specimens are preserved in Sir William J. Hooker's herbarium, and Dr. Harvey has figured them in his 'Phycologia Britannica,' but at the same time he expresses a doubt whether the plant should not be altogether erased from the list of species, and referred as a synonym to *C. Turneri*. Professor Agardh speaks of the species as unknown to him.

Callithamnion Daviesii. Davies's Callithamnion.

Fronds about a quarter of an inch high, growing in tiny tufts or continuously, irregularly branched; branches elongate, alternate, spreading; secondary branches short, alternate or secund, one or two only on each primary branch; branchlets short, secund, springing from the two or three lower joints of the secondary branches, forming what appear to be axillary tufts. Tetraspores tripartite, elliptical, borne on stalks on the axillary branchlets.

This species is parasitic on *Ceramium rubrum* and other sea-weeds, and, small as its tufts are, the axillary branchlets are so constantly infested with parasites and other extraneous matter that their tetraspores are frequently destroyed. It was discovered by the Rev. Hugh Davies on the Welsh coast, and is pretty generally distributed, though, from its small size, seldom found except by accident. It is annual, and in perfection in summer and autumn. The axillary position of the branchlets gives the typical form of this plant a very distinct appearance; but there are intermediate states which are not easy to determine.

Callithamnion virgatulum. The little twig Callithamnion.

Fronds about a quarter of an inch high, erect, growing continuously or in tiny tufts, much branched; branches long and straight, spreading, alternate or on one side only; branchlets consisting of a single joint, blunt, springing from every articulation of the primary and secondary branches, most frequently secund. Tetraspores formed from the branchlets, sessile or on short stalks.

The figure of this plant in Dr. Harvey's 'Phycologia Britannica' is sufficiently distinct from that of *C. Daviesii*, but it must be borne in mind that these represent extreme states of the two plants, and that there exists an almost complete series of intermediate forms. Dr. Harvey, "yielding to pressure from without," has reluctantly awarded specific rank to *C. virgatulum*, but Professor Agardh speaks of it only as a variety of *C. Daviesii*.

Callithamnion sparsum. The scattered Callithamnion.

Fronds about a quarter of an inch high, growing in minute

tufts, parasitic; branches few, simple, chiefly near the upper part of the frond; articulations about twice as long as broad. " Tetraspores obovate, sessile, mostly axillary."

This is another extremely minute and very doubtful species. Its simple fronds and flexuous branches are the characters relied on to separate it from *C. Daviesii* and *C. Rothii*, both of which it more or less resembles. The description of the tetraspores rests on the authority of Captain Carmichael.

GRASS-GREEN SEA OR FRESHWATER WEEDS.—CHLOROSPERMEÆ.

Fronds Grass-green, or in a few instances Purple or Olive, or very rarely Red. Propagation by simple cell division ; by the transformation of the colouring-matter of the cells into zoospores ; or, occasionally, by ordinary spores developed in proper spore-cases.

Comparatively few of the plants belonging to this division are marine. The remainder grow in freshwater streams, ponds and ditches, or even in damp places where there is no water. Their characters are not yet well understood, and in consequence many of the genera and species are but imperfectly determined. That portion only of the division which is marine will be described in this work.

Order XXI. SIPHONACEÆ.

Frond consisting of a single, filiform, branching cell, or of a sponge-like mass of many such cells interwoven; either naked or coated with carbonate of lime.

All the genera of this Order which are represented on our coasts belong to the section which is destitute of calcareous coating. They are very different from each other in external appearance, but their structure, when closely examined by the aid of a microscope, will be found to be similar.

Genus CIV. CODIUM.

Frond sponge-like, globular, cylindrical or flat, simple or branched, composed of interwoven, one-celled, branching threads, filled with green endochrome. Fructification, multitudes of minute zoospores contained in sporangia attached to the sides of the surface fibres of the frond.—CODIUM, from the Greek *kodion*, a hide.

This genus includes several species, some of which are very widely distributed, both in high and low latitudes in either hemisphere.

Codium bursa. The purse Codium.

Fronds spherical, hollow, composed wholly of slender threads closely interwoven, from the size of a pea to six or eight inches in diameter, growing several together, attached to the rock by matted fibres.

Only two or three English habitats, chiefly on the south coast, are recorded for this species. Curiously enough, the principal of these is Brighton, a locality not otherwise prolific of marine plants. I will not say it is abundant there, but I know that a magnificent specimen, six or eight inches in diameter, was recently obtained from thence. On the coast of Jersey many very fine specimens occur. They grow chiefly on rocks a little beyond the ordinary low-water mark, where they are unapproachable, except at the lowest spring-tides, and I have never had the good fortune to be in the island at that season. The plant is not, however, confined to these deep-water habitats, and I have several times found fair-sized specimens growing in shaded situations in pools near the very highest limit of the tide. I may add that its appearance in the water is similar to that of rounded pebbles, so that it may be easily overlooked, unless the sense of touch be used to aid the eye.

Codium tomentosum. The tomentose Codium.

Fronds growing several together or rarely singly, cylindrical or flattened, erect, forkedly branched, from a few inches to a foot or more long, and from a quarter to three-eighths of an inch in diameter, composed of an axis of numerous interwoven, slender fibres, from which issue horizontal, club-shaped filaments, whose tips constitute the surface of the frond. Zoospores in somewhat oval, nearly stalkless sporangia on the sides of the club-shaped filaments.

This plant is not only the most widely distributed species of the genus, but also rivals, in this respect, almost any other sea-weed: from the equator to either pole, and in both hemispheres it is abundant on nearly every coast. No less distinct than it is common, it may be readily recognized by its thick, sponge-like, dark-green branches, clothed with short, soft hairs, which, when spread out in water, give the plant a cotton-like appearance, whence it derives its name of *tomentosum*. It grows on rocks in pools, which are sufficiently deep to prevent the possibility of its being exposed to the air when the tide is down. It is perennial, and attains its most luxuriant growth in summer.

Codium amphibium. The amphibious Codium.

Fronds rising from an indefinite, spreading layer of en-

tangled fibres, growing several together but distinct from
each other, about half an inch high, erect, simple, cylindrical,
their axis composed of branched, interwoven, irregular
fibres, which throw off to the circumference club-shaped
filaments of the same nature, and nearly of the same form
as those of *C. tomentosum*.

This minute and very singular plant was discovered
by Mr. M'Calla, in 1843, on turf-banks, at extreme
high-water mark, near Roundstone, Galway, where it
was exposed alternately to the influence of salt and fresh
water, and occasionally to the absence of both.

Codium adhærens. The adhering Codium.

Frond spreading over rocks in irregular patches, from
one to two feet or more in diameter, composed of a layer
of entangled, interwoven, cylindrical fibres, from which issue
linear, club-shaped, vertical filaments; these are all of equal
length, and resemble in arrangement and appearance the
pile of velvet.

This species grows near low-water mark. It is perennial, and in perfection in summer. When wet it is of a
brilliant green colour, and its soft gelatinous texture
causes it to adhere closely to paper.

Genus CV. VAUCHERIA.

Fronds tufted, interwoven; each consisting of a single,
branched, one-celled, delicate, cylindrical filament, filled
with granular endochrome. Fructification, zoospores contained in sporangia attached to the sides of the branches
and accompanied by hooked cylindrical antheridia.—VAU-

CHERIA, named in honour of the Rev. M. Vaucher, a Genevese botanist.

Only three of the British species of this genus are marine, the remainder grow in fresh water. They are all insignificant, and cannot be easily preserved. The process of development and fertilization of their zoospores is, however, very curious, and has been carefully studied and described at great length by Pringsheim.

Vaucheria submarina. The submarine Vaucheria.

This species is figured and described in Berkeley's Gl. Br. Alg., p. 24, t. 8, and the same figure and description are given in the 'Phycologia Britannica.' The description is as follows :—" Plant growing in dense, fastigiate masses in muddy spots, covered by the sea at every tide. Threads far slenderer than in *Vaucheria dichotoma*, stained below by the mud, above dark-green, forked; the branchlets generally somewhat strangulated just above their insertion; the main stem clothed, above the part where the branchlet is given off, with numerous, almost sessile, more or less ovate, or lanceolate coniocystæ, which are pointed, at first entirely green, but eventually with a pellucid border. One single instance occurred in which the fruit consisted of two, placed end to end."

Vaucheria marina. The marine Vaucheria.

Fronds growing in tufts or singly, an inch or two high, sparingly, forkedly branched, slender. Fructification in pear-shaped, lateral sporangia.

This plant grows on mud, etc., between the tide-marks during summer, and is annual. When fresh it is of a bright grass-green colour; but this is changed to a glossy brown by the process of drying. It is reputedly rare, but this is most probably due to the fact that it is not easy to find its unobtrusive fronds, and that they are, therefore, frequently overlooked.

Genus CVI. BRYOPSIS.

Root fibrous. Fronds growing in tufts, erect, each consisting of a single, branched, one-celled, cylindrical filament, with membranaceous shining walls, and imbricated or pinnate branches and branchlets. Zoospores formed from a granular glutinous endochrome within the cell, from which they escape at maturity through apertures in its wall.—BRYOPSIS, from the Greek *bryon*, moss, and *opsis*, form.

Representatives of this genus are widely dispersed in all latitudes, and in consequence many species have been described. It is, however, doubtful whether some at least of these are not merely varieties of *B. plumosa*, changed more or less in form by the circumstances of their growth. Dr. Harvey, in his 'Nereis Boreali-Americana,' expresses his belief that they are, and adduces strong evidence in support of his opinion.

Bryopsis plumosa. The feathery Bryopsis.

Fronds growing in tufts or singly, from two to four inches high, once or twice pinnate, the lower part of the stem, of all the divisions of the frond, naked.

The light feathery form of the fronds of this species, and their bright green colour, which is preserved, and becomes lustrous when they are dried, entitle it to rank among the most beautiful of the green series of sea-weeds. It is annual, and may be found in shore-pools, and at greater depths on most parts of our coasts during summer and autumn.

Bryopsis hypnoides. The Hypnum-like Bryopsis.

Fronds growing in tufts, from four to six inches high, much branched; branches long, rod-like, set on all round the stem, either simple or bearing a second similar series; branchlets irregularly pinnate, very slender and delicate, generally confined to the upper part of the branches.

There is some general resemblance between this species and the last, and there are exceptional forms of each, whose identity it is not always easy to determine. The distinctive characters of *B. hypnoides* are its numerous long branches, and its soft silky texture. This species is annual, and flourishes in summer. It grows on rocks or sea-weeds in shady pools of various depths, and is pretty generally distributed round our coasts.

Order XXII. ULVACEÆ.

Frond of a Grass-green or Purple colour, composed of many minute cells, forming a thin membrane of indefinite shape, flat or tubular, simple or branched. Fructification, ciliated zoospores developed in the cells of the frond.

This Order affords an illustration of the natural alliance of sea-weeds of different colours. Most of the genera composing it are of the bright, pale green colour, which characterizes the division; but two, *Porphyra* and *Bangia*, are purple. Some writers have separated these from the remainder, and placed them in the Red division; but this course has not been sanctioned by the best authorities, who have considered that structure and mode of fructification are characters of greater value than colour.

Genus CVII. PORPHYRA.

Frond a flat, irregularly-shaped, very delicate purple membrane. Fructification, purple spores, arranged in groups of four throughout the frond.—PORPHYRA, from the Greek *porphureos*, purple.

There is a wide divergence of opinion as to the number of species into which this genus should be divided. Kützing has described sixteen; but other writers treat many of these as only varieties. Hitherto two at least of the forms found on our coast have been considered to be distinct; but Dr. Harvey's most recently expressed opinion is that they are not, and that "if we contend for two species, with equal justice we might make half-a-dozen." I think so too, and I therefore unite both forms under one name.

Porphyra vulgaris. The common Porphyra.

Fronds of very variable size and shape, some ribbon-like, and attached at the base, others spreading from a central root, and more or less divided into irregular, ragged segments; their usual colour is a dark purple, which is derived from the fruit-bearing cells, that extend over the whole fronds. Specimens not in fruit are of a blackish-green hue.

The different forms of this species are all, I believe, the results of variations of season and situation. In early winter, and near high-water mark, the narrow, minutely midribbed form (*P. linearis* of Greville) will be found. This is succeeded, often in the same locality, by the broader, waved fronds of the form which has been hitherto considered the type of this species. These again, in deeper water, and as the season advances, pass by scarcely perceptible degrees through several intermediate states into the extremely expanded, divided plant that is known as *P. laciniata*. In almost every locality, generally near the influx of some freshwater stream, the filmy, fragile fronds of this species are to be found, floating helplessly in the eddies caused by the advancing or receding waves, or spreading themselves in purple, shining patches over rocks, stones, or mud when the tide is out. As distinct as they are common, there can be no doubt as to their identity, for there is no other British sea-weed that they at all resemble. They are, however, difficult to preserve, for they do not, as a rule, adhere closely to paper, and require heavy and long-continued pressure and careful drying. Even after the most skilful and patient treatment they will hardly bear a momentary exposure to the air without beginning to curl up and crack. Laver, sloke, or sloukawn, as it is variously called in England, Scotland, and Ireland, is made by boiling the fronds of this species for several hours. It is afterwards fried, and eaten with vinegar and pepper; and although by no means a tempting dish to the eye, it is, I believe, both palatable and wholesome.

Genus CVIII. **BANGIA**.

Fronds hair-like, composed of numerous cellules, which radiate from a central cavity, and are enclosed in a con-

tinuous, transparent sheath. Fructification, purple zoospores, one of which is formed in each cell of the frond.—
BANGIA, named in honour of Hoffman Bang, a Danish botanist.

Some of the species of this genus are marine, some fluviatile, and others grow indifferently in either salt or fresh water. In this respect, as well as in the tubular form of their fronds, they resemble the *Enteromorphæ*, from which they differ in colour. It would, indeed, appear that the purple, tubular *Bangia* bears the same relation to the purple, flat *Porphyra*, that the green, tubular *Enteromorpha* does to the green, flat *Ulva*.

Bangia fusco-purpurea. The brown-purple Bangia.

Fronds from two to three inches long, slender, simple, straight, growing in decumbent, silky masses, which float freely in the water, of a dark purple colour.

This species grows on rocks or submerged wood, in salt or fresh water. Its characters vary a good deal, according to the age of the specimens. It is an interesting object under the microscope.

Bangia ciliaris. The fringe-like Bangia.

Fronds parasitic on Algæ, very minute, "scarcely the tenth of an inch long."

Writing of this plant in his 'Nereis Boreali-Americana,' Dr. Harvey says, "Possibly it may be only the very young state of *B. fusco-purpurea*; but the habitat is different, and the colour much brighter."

Bangia ceramicola. The Ceramium Bangia.

Fronds parasitic on Algæ, about an inch and a half long, simple, very slender, and flaccid, of a beautiful rose-red colour.

When examined under a microscope, the fronds of this species appear to be jointed, and in mature specimens the spaces between the joints to be striped lengthwise. These appearances are deceptive, and in reality what seem to be joints are but constrictions of the transparent sheath of the frond, and the stripes are produced by the division of the granular endochrome.

Bangia elegans. The elegant Bangia.

Fronds parasitic on Algæ, minute, forkedly branched, with spreading axils.

This species is very rare. The only recorded British specimens are from Portaferry, where they were dredged many years since by Mr. Thompson.

Genus CIX. ENTEROMORPHA.

Fronds green, tubular, simple or branched, composed of a net-like membrane. Fructification, zoospores produced in the cells of the frond, generally in groups of four.— ENTEROMORPHA, from the Greek *enteron*, an entrail, and *morphe*, form.

The plants included in this genus are probably more widely distributed, and grow in greater variety of circumstances, than those of any other sea or water-weed. Not content with ranging over the ocean, from the poles to the equator, they penetrate inland, and may be found in rivers, brooks, and even ditches.

Enteromorpha cornu-copiæ. The cornu-copiæ Enteromorpha.

Fronds about an inch high, stalked, tubular, of very small diameter at the base, becoming suddenly wider above, and eventually bursting into a miniature goblet, usually parasitic on *Corallinæ*, etc. Fructification dispersed throughout the frond.

I have hesitated to degrade this plant to the rank of a variety, as it has been admitted as a species by authors who have had better opportunities to study it than I have. At the same time, I am inclined to agree with those botanists who regard it as only a worn state of a stunted form of *E. intestinalis*.

Enteromorpha intestinalis. The intestine Enteromorpha.

Frond a simple, elongate, membranous sac, taper at the base, obtuse at the tip, and of very variable length and diameter. Fructification in the cells of the membrane of which the frond is composed.

This plant is found on all parts of our coasts, in tidal and other rivers, in canals, and even in ditches. It appears to grow indifferently in salt and fresh water, and to attain the greatest luxuriance in situations where the two mingle. The fronds vary considerably in size, both in length and breadth. They are occasionally constricted or crisped, and their inflation is more or less irregular. With these exceptions the character of the plant is very constant, and its specific identity may be always readily determined by the total absence of branches of any description. The Japanese use this weed to thicken soups, much as vermicelli is used in this country. Probably the substance which is occasionally imported from China, under the name of "artificial bird's-nest," and which is, as I can testify from personal experience, applied with excellent effect to the same purpose, has a like, or at least a kindred origin.

Enteromorpha compressa. The compressed Enteromorpha.

Frond branched, elongate, tubular or somewhat compressed; branches long, simple, gradually tapered at the base, obtuse at the tip. Fructification in the cells of the membrane of which the frond is composed.

In nearly every latitude this plant abounds, both on the seacoast and in the estuaries of tidal rivers. It grows at all seasons on rocks, shells, and woodwork, and varies very much in size. It frequently covers a vast extent of the surface of perpendicular rocks, and marks the level of high tide. In such situations its fronds are numerous, densely packed, and small, like stunted, very bright green grass. In the shore pools it is more fully developed, and at those points where fresh water flows into the sea, its fronds reach their maximum luxuriance, attain considerable length and breadth, and become inflated.

Enteromorpha clathrata. The latticed Enteromorpha.

Frond much branched, slender, tubular, cylindrical; branches more or less spreading, sometimes squarrose, beset with simple, slender, spine-like, awl-shaped branchlets; cells of the membrane rectangular. Fructification contained in the cells.

This is a very generally distributed and variable plant. It is less robust and more branched than *E. compressa*, and all the divisions of its fronds are acute at the tip, while those of the latter species are obtuse. No less than three forms, besides the type, have been figured and described by authors as distinct species; but the result of recent research has been to degrade them to the rank of varieties, and as such I insert them. They are—

E. clathrata var. Linkiana. Link's Enteromorpha,

Described by Dr. Greville, and figured by Dr. Harvey (Phyc. Brit. tab. 344), from a single specimen collected by Captain Carmichael, and preserved in the Dublin University Herbarium.

E. clathrata var. erecta. The erect Enteromorpha,

Which appears to be a luxuriant deep-water form, with longer, more slender, less rigid, and more numerous branches and branchlets than the typical plant (Phyc. Brit. tab. 43).

E. clathrata var. ramulosa. The sharp-branched Enteromorpha,

Which is distinguished by its squarrose habit, curved, entangled branches, and numerous short, horizontally spreading, spine-like branchlets (Phyc. Brit. tab. 245).

Enteromorpha Hopkirkii. Hopkirk's Enteromorpha.

Frond from a few inches to a foot long, much branched, very slender, soft in texture; branches erect, alternate or opposite, repeatedly divided, tapered; branchlets awl-shaped, minute, composed of a single series of cells, and consequently jointed. Fructification, granules of endochrome in the centre of large transparent cells.

The recorded habitats of this species are not numerous. It was first found at Torbay by Mr. Griffiths, and subsequently at Carrickfergus by Mr. M'Calla. It also occurs on the coast of North America; but does not appear to attain so large a size there as in this country. Its distinctive character is the large size of the cells of which its fronds are composed.

Enteromorpha percursa. The spreading Enteromorpha.

Fronds hair-like, simple or rarely branched, growing in tangled masses, composed of small square cells, which are nearly filled with endochrome.

This is an insignificant species, whose identity is not by any means satisfactorily determined.

Enteromorpha Ralfsii. Ralfs's Enteromorpha.

Fronds hair-like, simple or rarely branched, long and slender, composed of large transparent cells, each of which contains a granule of bright green endochrome.

This is another doubtful species, and the information concerning it is meagre and confused.

Genus CX. ULVA.

Frond a flat, irregularly-shaped, leaf-like, green membrane. Fructification, green spores, generally in groups of four, scattered throughout the frond.—ULVA, from the Celtic *ul*, water.

This genus, as hitherto known to British botanists, has been variously rearranged. As regards our native species, it may be most conveniently divided into two sections. One, which corresponds for the most part with the genus *Phycoseris* of Kützing, has the frond composed of a double layer of cellules, or rather of two separable but closely cohering membranes. The other, *Ulva* of Kützing, has only a single membrane or layer of cellules. Believing it to be desirable to avoid as much as possible all changes of names, I follow the example of Dr. Harvey in his 'Nereis Boreali-Americana,' and retain the name of *Ulva* for both sections.

Ulva Linza. The narrow Ulva.

Frond composed of two layers of cellules, from a few inches to more than a foot long, and from less than half an inch to two inches wide, linear-lance-shaped, taper at the base, more or less acute at the tip, waved at the margin, of thin, membranaceous substance, and a bright grass-green colour. Fructification dispersed over the whole frond.

This species is annual, and grows during summer on rocks and stones about midway between the tide-marks. It is not very common, but may be found in many localities round our coast. The outline of the frond is more regular than that of the fronds of any other British *Ulva*, and the plant, particularly when seen growing under water, is very graceful and beautiful. Dried specimens preserve their colour well, and adhere closely to paper; but they are very susceptible of injury from exposure to the air, or to a sudden change of temperature. The greatest care must, therefore, be used both during the process of preparation for the herbarium, and subsequently, whenever they may be examined, or they will crack and curl up from the paper and be spoiled. These remarks apply with equal, or greater, force to dried specimens of all the species of this genus.

Ulva latissima. The very broad Ulva.

Frond composed of two layers of cellules, from a few inches to two feet long, and about half as wide, extremely variable in size and shape, with an irregular outline, and a ragged, waved or sinuate margin, of very thin substance, and soft, but moderately tough, when young of a vivid, somewhat bluish green colour, becoming paler with age. Fructification dispersed throughout the frond.

Ranging from between the tide-marks into ten or twelve fathoms of water, and from the Equator to the Arctic and Antarctic circles, this plant appears to be everywhere abundant, and to flourish almost equally in warm and cold climates. It is to be found in summer and autumn, and is annual. The fronds grow in tufts, and are always flat; they do not adhere to paper. Mature specimens of this plant are frequently more or less covered with minute, dot-like tufts of *Myrionema*.

Ulva lactuca. The lettuce Ulva.

Frond composed of a single membrane, or layer of cellules, from two to six inches high, at first a pear-shaped sac, which soon bursts into irregular segments, very thin and delicate, of a pale yellowish-green colour, glossy, and adhering closely to paper when dry. Fructification dispersed throughout the frond.

This species is distinguished from the preceding by its fronds being composed of a single membrane, and according to the arrangement to which I have alluded, would be the only British representative of the genus *Ulva*. It grows on rocks, shells, and small algæ between the tide-marks in spring and in early summer, and is annual. It is found in many localities round our coasts, but is less common than either of the allied species.

Order XXIII. CONFERVACEÆ.

Frond green, thread-like, composed of cylindrical cells, which are usually of greater length than their diameter, and are joined end to end. Fructification, minute, ciliated zoospores formed from the colouring matter of the cells.

A very large number of genera and species are included in this order. Some of them are marine, and others grow in brackish or in fresh water. It is very difficult to describe them, or to determine with certainty their generic and specific characters, and it is probable that many which have been considered to be distinct are in reality but different forms of the same plant.

Genus CXI. LEPTOCYSTEA.

Frond tufted, erect, jointed, branched; cells elongate, only one between each two sets of branches, filled with granular endochrome, from which the zoospores are developed.—LEPTOCYSTEA, from the Greek *leptos*, slender, and *kustis*, a bladder.

The only British species belonging to this genus has been hitherto included among *Cladophoræ*, and I have hesitated whether to depart from that arrangement. There can, I think, be little doubt that the genus *Cladophora*, containing, as it has been made to do, all the branched species of the order *Confervaceæ*, must be ultimately divided; and it appears to me that this plant possesses sufficiently distinct characters to entitle it to be placed in a separate genus. This course has been adopted by Dr. Gray in his 'Handbook of British Water-weeds,' and I follow it here.

Leptocystea pellucida. The transparent Leptocystea.

Fronds from three to eight inches high, growing in tufts or singly, very rigid; stem undivided below, forked above, and ultimately repeatedly branched in a dichotomous or trichotomous manner; branches opposite, or more or less whorled, springing regularly from each articulation.

This is a very handsome, and by no means uncommon plant. It grows in pools near low-water mark, and is probably annual. The arrangement of the branches at the top of each of the long one-celled joints of the frond is a very obvious character, and affords a never-failing means of recognition.

Genus CXII. **CLADOPHORA.**

Frond tufted, uniform, jointed, branched; cells oblong, usually more than one between each two branches or sets of branches, filled with granular endochrome, from which the zoospores are developed.—CLADOPHORA, from the Greek *klados*, a branch, and *phoreo*, to bear.

A very large number of species are included in this genus: some of them are marine, and others grow in brackish or in fresh water. Their specific characters are not well defined, and even the limits of genus have yet to be authoritatively traced. Several modes of grouping the species into new genera have been proposed; but none of these are sufficiently established to be admissible into a popular work. I have, therefore, retained the name *Cladophora*, and contented myself with arranging the species according to the most recent and best-defined systems.

Cladophora rupestris. The rock Cladophora.

Fronds densely tufted, rigid, shrub-like, from three to nine inches long, of a dark-green colour; branches opposite; branchlets awl-shaped; cells about three times the length of their diameter.

This species is annual, and grows during summer and autumn on rocks between the tide-marks, and in deep water. It is common all round the British coast, and may be readily recognised by the hard, rigid texture and dark-green colour of its fronds. Dr. Harvey writes—" The process of cell-division is well illustrated in this species, and may be observed even in dried specimens, so perfectly does the endochrome recover its form. The cells of the middle portions of the branches divide, as well as those of the younger ramuli, and consequently consecutive cells are found of various lengths."

Cladophora rupestris, var. β. distorta.

"Tufts rooting in the mud, depressed; filaments short, much curled and matted together; ramuli squarrose."—*Harv.*

"Found on sub-marine peat at Biturbui Bay, Connemara, by Mr. M'Calla."

Cladophora rectangularis. The rectangular Cladophora.

Fronds tufted, rigid, entangled, from six inches to a foot long; branches opposite, spreading, distant, pinnate; branchlets opposite; cells about twice the length of their diameter, uniform throughout the frond.

This very beautiful and rare species was first found at Torquay, by Mr. Borrer, in 1832. It grows in deep water during summer, and is annual.

Cladophora lætevirens. The pale-green Cladophora.

Fronds tufted, much divided, from six to eight inches long, of a soft texture and yellow-green colour; branches spreading, crowded, of unequal lengths; branchlets somewhat curved, obtuse at the tip; cells of the branches about six times the length of their diameter, those of the branchlets about thrice.

This species is very common. It grows on rocks and sea-weeds, between the tide-marks, during summer, and is annual. It resembles the fresh-water *C. glomerata*, and the two plants are perhaps only different forms of the same species, altered by the circumstances of their growth, the one being distinctly marine, and the other as distinctly fluviatile.

Cladophora diffusa. The diffused Cladophora.

Fronds growing in lax tufts, rigid, bristly, much branched, from a few inches to a foot long, of a dark-green colour when young, becoming paler and yellowish when mature; branches long, distant, irregular, alternate or forked; branchlets few, simple, secund, confined to the upper divisions of the branches; cells about three times the length of their diameter, uniform throughout the frond.

This species grows on rocks and in pools, between the tide-marks, and is a summer annual. It resembles *C. Hutchinsiæ*, and probably these two forms are only different states of the same plant. I am inclined to believe that both are specifically identical with *C. lætevirens*.

Cladophora Hutchinsiæ. Miss Hutchins's Cladophora.

Fronds growing in tufts, rigid, bristly, much branched, from a few inches to a foot long, of a glaucous green colour and crisp texture; branches long, curved or zig-zag, distant, irregular, alternate or forked; branchlets few, alternate or secund, simple, or with short shoots on their inner sides, very blunt at the tip; cells about twice the length of their diameter; joints of the frond constricted.

This plant grows on rocks in deep tide-pools during summer. It is annual, and rare.

Cladophora Macallana. M'Calla's Cladophora.

Fronds rigid, bristly, brittle, much branched, rolled together in bundles, from a few inches to a foot or more long, of a bright grass-green colour; branches bent, somewhat zig-zag, of unequal lengths, irregularly disposed; branchlets alternate or secund, short, spreading, recurved, blunt at the tip; cells about twice the length of their diameter.

This species grows in deep water during summer, and is annual. It resembles *C. rectangularis* in general character and appearance, but the branchlets of that plant are opposite, while those of *C. Macallana* are alternate or secund.

Cladophora falcata. The sickle-shaped Cladophora.

Fronds growing in dense tufts, rigid, slightly interwoven, curved, from three to four inches high, of a bright darkgreen colour and crisp texture, much branched; branches repeatedly divided, sickle-shaped; branchlets short, secund, obtuse; cells about three times the length of their diameter, nearly uniform throughout the frond, transparent at the margin.

This very elegant *Cladophora* was first gathered by Dr. Harvey, on the coast of Kerry, in 1845, and subsequently by Miss White in Jersey. It grows in rockpools near low-water mark, and is annual. The curvature of the frond is the distinctive character of this plant, and it is a little doubtful whether that be a sufficient title to constitute a valid claim to specific rank. In other respects it resembles *C. lætevirens*.

Cladophora glaucescens. The glaucous Cladodophora.

Fronds growing in tufts, very slender, much branched, from two to five inches long, of a glaucous green colour, and softish texture; branches somewhat zig-zag in the main divisions, straight and erect in the lesser; branchlets long, erect, secund; cells about three times the length of their diameter, uniform throughout the frond.

This species grows between the tide-marks in summer, and is annual.

Cladophora flexuosa. The flexuous Cladophora.

Fronds growing in tufts, slender, much branched, from a few inches to half a foot long, of a pale green colour, and softish texture; branches somewhat zig-zag, long, of unequal length, the secondary series simple, or set with short, comb-like branchlets on their outer sides; cells from twice to four times the length of their diameter, those in the lower part of the frond being the shortest.

This plant was first found by Mrs. Griffiths in tide-pools at Torquay. It closely resembles *C. glaucescens*, and is probably only a variety of that species.

Cladophora refracta. The reflexed Cladophora.

Fronds growing in dense tufts, about three inches high, slender, somewhat rigid, of a glossy yellow-green colour, and rather harsh texture; main stems rolled into rope-like bundles; branches free, spreading, much divided; branchlets short, recurved, set with short, comb-like, secund ramelli; cells about twice the length of their diameter.

This species grows in rock-pools near low-water mark, and is a summer annual.

Cladophora albida. The whitish Cladophora.

Fronds densely tufted, very slender, entangled into rope-like bundles at the base, from a few inches to a foot long, of a soft silky texture, and pale yellow-green colour, becoming whitish when dry; branches crowded, mostly opposite, of various lengths, those of the upper part of the frond spreading; branchlets opposite, or secund ; cells about four times the length of their diameter.

This species is common on the south and west coasts of England and Ireland. It grows on rocks and seaweeds near low-water mark during summer, and is annual. The distinctive characters are that the threads of the fronds are very long, delicate, and soft. In other respects it resembles *C. refracta.*

Cladophora gracilis. The slender Cladophora.

Fronds growing in long, tangled tufts, from six inches to a foot in length, of a yellow-green colour and silky texture, much branched; branches alternate, twisted, much divided; branchlets long, slender, secund, comb-like, tapering from the base to the tip; cells about four times the length of their diameter, nearly uniform throughout the frond.

This plant grows on *Zostera*, etc., in deep water, and is a summer annual. Its characters are better defined than those of most of its allies ; but it is, unfortunately, rare, and confined to certain favoured localities.

Cladophora Brownii. Brown's Cladophora.

Fronds growing in dense tufts, erect, rigid, from half an inch to an inch high, inextricably interwoven, slightly branched ; branches long, nearly simple; axils acute; cells about four times as long as broad, those below swollen at the top, those above cylindrical.

This plant was first discovered and described by that eminent botanist, the late Robert Brown, after whom it has been named by Dr. Harvey. It grows on rocks in caves, etc., where it is exposed to both salt and fresh water. It is perennial, and rare.

Cladophora Magdalenæ. Miss Magdalene Turner's Cladophora.

Fronds very slender, interwoven, slightly branched, about an inch long, of a dull dark-green colour and rigid substance; branches irregular, alternate or forked, spreading, curved; branchlets scattered, secund, sickle-shaped, not always present; cells about four times as long, as they are broad, those above rather shorter than those below.

This insignificant plant was discovered growing on the roots of algæ, in Jersey, by the lady whose name it bears. It is doubtful whether it be entitled to specific rank.

Cladophora flavescens. The yellowish Cladophora.

Fronds growing in floating masses, very slender, interwoven, sparingly branched, of a yellowish-green or yellow colour, and soft silky texture; branches forked, spreading, angularly bent; branchlets few, alternate or secund; cells about eight times the length of their diameter.

This species is annual, and grows, during summer, on the surface of pools of brackish or fresh water. It may be known by the yellow colour and long cells of its fronds.

Cladophora nuda. The bare Cladophora.

Fronds growing in lax tufts, about three inches high, rigid, slender, straight, forked, of a dull-green colour, slightly branched; branches distant, erect, almost bare; branchlets few, robust, blunt at the tip, set on to the branches at a very acute angle; cells very long, with wide, transparent margins.

This is a doubtful species founded on specimens gathered long since "on basalt rocks, between tide-marks, at Port Stewart, co. Antrim, by Mr. D. Moore."

Cladophora Balliana. Miss Ball's Cladophora.

Fronds growing in tufts, about nine inches long, very slender, of a grass-green colour and soft texture, much branched; branches much divided in an alternate manner, intricate, somewhat virgate; branchlets very slender, secund, tapering at the tip, generally one- or two-celled; cells many times the length of their diameter, those above shorter than those below, all filled with dense granular endochrome, and surrounded by a broad transparent border.

This species was discovered at Clontarf by Miss Ball, and was considered by Dr. Harvey to be distinct from all British *Cladophoræ*.

Cladophora fracta. The broken Cladophora.

Fronds growing in tufts, at first attached, then floating free, in masses of considerable size, entangled, rather rigid, from six inches to a foot long; branches distant, several times forkedly divided, spreading; branchlets few, alternate or secund, not always present; axils of the branches very wide; cells from three to six times the length of their diameter, when young cylindrical, becoming elliptical as the zoospores are developed.

This species is placed by Dr. Gray in a new genus, under the name *Vagabundia*, characterized as follows: —" Frond green, forming tufts, branched; cells elongate, subcylindrical, the cells of the stem becoming swollen and converted into sporangia." *C. fracta* grows in brackish or fresh water, and is common. In the earlier stages of its growth it is of a grass-green colour, but becomes darker as it approaches maturity. It does not adhere closely to paper.

Cladophora arcta. The straight Cladophora.

Fronds from two to four inches long, growing in dense tufts, at first only slightly tangled at the base, eventually matted into a sponge-like, circular mass; branches numerous, erect, straight, furnished below with root-like fibres; branchlets opposite or secund, blunt at the tip, their axils very acute; cells from one to ten or twelve times the length of their diameter, the increase of length being gradual from the base to the tip of the frond.

This and the two following species are placed by Kützing in his section *Spongomorpha*, and Dr. Gray has founded a new genus, *Spongiomorpha*, in which he has associated them with *C. Gattyæ*, and I am inclined to think that *C. repens* might be also included, as it has the root-like fibres, and is otherwise similar. Dr. Gray gives the characters of *Spongiomorpha* as follows:—" Frond erect, forming tufts; branches virgate, dichotomous, subsimple, with root-like fibres; cells oblong, filled with granular endochrome, becoming zoospores, marine."

C. arcta grows on rocks in tide-pools, during the greater

part of the year, and varies much in appearance at different seasons. When young the colour is a bright-green, the texture is silky, and the plant adheres closely to paper; when fully developed the colour becomes dull, the texture coarse, and the adherent quality is lost. In the latter state there is generally a growth of young branches at the tips of the fronds, which forms a bright margin to the tuft, and contrasts strikingly with the duller colour of the remainder.

Cladophora lanosa. The woolly Cladophora.

Fronds about an inch long, growing in dense, sponge-like, level-topped, globular tufts, very slender, interwoven; branches straight, rod-like, spreading, generally alternate, furnished with root-like fibres; branchlets few, scattered; axils acute; all the divisions of the frond blunt at the tip; cells of the lower part of the frond two or three times, those of the upper part about six times the length of their diameter; colour a pale yellow-green, which becomes almost white when the plant is dry.

This species grows on rocks, sea-weeds, and *Zostera* at various depths, and is common in most localities, attaining the greatest luxuriance and abundance in cold regions. It very closely resembles *C. arcta*, but is smaller, more slender and less branched. Specimens should be laid out in salt water, as the cells are liable to burst and discharge their endochrome if the fronds be placed in freshwater. For the same reason they should not be subjected to very great pressure while drying. They adhere to paper, but do not retain their gloss.

Cladophora uncialis. The inch Cladophora.

Fronds slender, from one to two inches high, growing, from a common base, in dense, sponge-like, interwoven, compact or divided, somewhat level-topped tufts; branches distant, spreading, alternate or secund, set with root-like, jointed fibres; branchlets few, simple, curved; axils rather wide; cells about twice the length of their diameter, nearly uniform throughout the frond; colour, of a growing plant, bright-green, of a dried specimen, pale yellow-green.

This species grows on rocks near low-water mark, and is pretty generally distributed round our coasts from Orkney to the Channel Islands. It is very closely allied to *C. lanosa,* but the tufts are more intimately interwoven, and less distinctly level-topped; the cells are uniform throughout the frond; the root-like fibres are more numerous; and *C. uncialis* is usually found growing on rocks, while *C. lanosa* is almost constantly parasitic. This species should also be laid out in salt water, in order to preserve the endochrome.

Cladophora Gattyæ. Mrs. Gatty's Cladophora.

" Filaments about an inch long, as thick as human hair, or somewhat thicker, matted together in dense ropy tufts, irregularly branched, somewhat dichotomous, the angles rounded; ramuli few and patent; articulations very uniform, about once and a half as long as broad; filled with olivaceous (?) or dull green endochrome, and separated by exceedingly narrow dissepiments; apices on my specimens often broken; substance membranaceous, adhering to paper."—*Harvey.*

This species is named in honour of the talented authoress, who collected the specimens on which it was founded. I am not acquainted with the plant, and I am therefore unable to add anything to Dr. Harvey's description, which I have given above.

Cladophora repens. The creeping Cladophora.

Fronds very slender, the lower part creeping, set with root-like fibres, which connect them into very dense, circular or oblong, cushion-like tufts of an inch or more in diameter; branches springing from the joints of the main thread at a right angle to the decumbent portion, about half an inch long, erect, simple or once forked; branchlets few, simple, secund; cells very long, only three or four in each filament; colour when growing, dark-green, when dried, dingy olive-green.

The first British specimen of this species was collected in Jersey, by Miss Turner. It was cast on shore, and the plant has not been again observed in the same locality. As it probably grows in deep water, and is very small, this fact is by no means remarkable, and certainly must not be accepted as a proof that specimens do not exist in the neighbourhood. The other recorded habitats are the Mediterranean and the Adriatic, so that the species would appear to be a native of the South. The very great length of the cells affords a ready means of distinguishing *C. repens* from all British *Cladophoræ* with which it can be confounded. It is furnished with root-like fibres, and for that reason I place it near those species which possess that character.

Cladophora Rudolphiana. Rudolphi's Cladophora.

Fronds from a few inches to more than a foot long, very slender, much branched, interwoven into flaccid, somewhat gelatinous tufts of considerable length; branches twice or thrice forked or alternate, irregular; branchlets long, comblike, secund, gradually tapered towards the tip; cells many times longer than broad; those of the upper part of the frond being shorter than those of the main stem, which are sometimes swollen.

This species grows on *Zostera* and sea-weeds in deep water during summer, and is annual. It is very abundant in certain localities on the coast of Ireland, and has been found on other parts of our shores and in the Adriatic. It adheres closely to paper, and preserves its brilliant glossy green colour when dried. Some of the cells of this plant are swollen in the middle, and on this character Dr. Gray has founded a new genus, *Cystothrix*, which he describes as follows :—"Frond erect, dichotomously branched; cells cylindrical, with a swollen one before the furcation of the branchlets, filled with endochrome, marine."

Genus CXIII. CHÆTOMORPHA.

Frond a simple, hair-like, membranaceous or cartilaginous filament, formed of oblong cells filled with granular endochrome, the bottom cell larger than the others.— CHÆTOMORPHA, from the Greek *chaite*, flowing hair, and *morphe*, form.

This genus was founded by Kützing for the reception of a portion of the species formerly included in the genus *Conferva*. Some of the species of which it is composed are marine, and others fluviatile. They are all comparatively insignificant, and their characters are difficult to determine, and still more difficult to describe in intelligible language.

Chætomorpha Melagonium. The dark-green Chætomorpha.

Root shield-shaped. Threads robust, bristle-like, erect, rigid, growing singly or several together; cells about twice the length of their diameter.

This species grows on rocks in tide-pools near low-water mark, and is perennial. It is widely distributed, and may be readily distinguished by the thickness and wiry texture of its filaments. It was formerly *Conferva Melagonium*.

Chætomorpha ærea. The pale-green Chætomorpha.

Root shield-shaped. Threads bristle-like, straight, rigid, brittle, growing in tufts; cells about the length of their diameter.

This species grows on sand-covered rocks between the tide-marks, and is common on the British coast. It is less rigid, more slender, and of a much paler colour than *C. Melagonium*. It was formerly a *Conferva*.

Chætomorpha sutoria. The bristle Chætomorpha.

Threads from a few inches to a foot or more long, bristle-like, curved, growing, unattached and floating, in large, loosely interwoven masses; cells nearly twice the length of their diameter, dividing transversely in the middle at maturity.

This species was also formerly a *Conferva*. It grows in ditches and pools, which are rendered more or less brackish by the influx of sea-water. It is of a dark-green colour, and rigid texture, and does not adhere to paper when dry.

Chætomorpha linum. The twine Chætomorpha.

Threads robust, from a few inches to several feet long, curled, rigid, growing, unattached and floating, in loosely interwoven layers of considerable extent; cells about the length of their diameter, dividing transversely at maturity.

This species varies much in colour, from pale to dark-green, and is sometimes variegated. It is of a rigid brittle texture, and does not adhere to paper when dry. It grows in large quantities in tidal ditches. This, too, was a *Conferva*.

Chætomorpha tortuosa. The tortuous Chætomorpha.

Threads slender, tortuous, rigid, curled, growing in closely interwoven layers, which spread over and adhere to projections of rock, and other substances with which they may come in contact; cells varying in length, usually about twice that of their diameter.

This plant grows on rocks, etc., in tide-pools, and is not uncommon. It was formerly a *Conferva*.

Chætomorpha implexa. The interwoven Chætomorpha.

Threads very slender, limp, curled, intimately interwoven, growing in flattish masses on rocks and sea-weeds; cells rather longer than they are broad.

This species was also formerly a *Conferva*. It is of a bright grass-green colour and soft texture.

Chætomorpha arenicola. The sand-inhabiting Chætomorpha.

"Threads soft, simple, extremely fine, matted, somewhat crisp, at first uniform pale-green, at length distinctly jointed; articulations once and a half as long as broad, dotted; interstices pellucid."—*Berkeley*.

This plant appears to be but little known, and is not referred to by Kützing. The above is Mr. Berkeley's description of it, and he further speaks of it as "creeping on the sandy margins of pools in a salt-marsh, periodically flooded, forming a thin, soft, delicate, crisped web, of a pale yellow-green."

Chætomorpha arenosa. The sand Chætomorpha.

Threads slender, nearly straight, rigid, harsh, about six inches long; cells about four times the length of their diameter.

Quoting from Captain Carmichael, Dr. Harvey writes: —"This species occurs in fleeces a yard or more in extent, and of a peculiar structure. They consist of several thin layers, placed over each other, but so slightly connected that they may be separated like folds of gauze, to the extent of many inches, without the least laceration." *C. arenosa* grows on sandy shores between the tide-marks in Scotland and Ireland. It differs from the other species of the genus in the greater length of its cells.

Genus CXIV. CYTOPHORA.

"Filaments membranaceous, interwoven, attached or floating; cells oblong, with here and there a pair of swollen discoloured cells." (*Dr. J. E. Gray*).—CYTOPHORA, from the Greek *kutos*, a cavity, and *phero*, to bear.

This genus contains a single British species, which was formerly *Conferva litorea*.

Cytophora litorea. The shore Cytophora.

Threads robust, rigid, crisped, about four inches long, sometimes angularly bent, growing in loosely interwoven layers of considerable extent, which either float or spread themselves over mud, etc.; cells about once and a half as long as broad. At irregular intervals of the thread a pair of cells, larger than the others, unite into a spindle-shaped knot, in which a dark-coloured endochrome is developed.

This plant grows on muddy shores between the tide-marks, in the estuaries of rivers, and in salt-water ditches. It is a summer annual.

Genus CXV. HORMOTRICHUM.

Frond a simple or slightly branched, hair-like, gelatinously membranaceous filament, attached by its base, and formed of very short cells, which have thick, soft walls, and are filled with green, granular endochrome, whence darker-green sporidia are developed; nodes constricted.—HORMOTRICHUM, from the Greek *hormos*, a necklace, and *trichoma*, hair.

All the species of this genus are marine, and were formerly included in the genera *Conferva* and *Lyngbya*. They differ from the *Chætomorphæ* in habit and texture, and in the changes that take place in the endochrome. They all grow on rocks, etc., between the tide-marks.

Hormotrichum Younganum. Young's Hormotrichum.

Threads from one to three inches long, about as thick as human hair, erect, straight or curved, of a bright green colour, not glossy, and but slightly gelatinous, growing in tufts on rocks, etc.; cells varying in length from rather shorter than they are broad to nearly twice as long; those of young threads which are increasing in length are quite filled with endochrome, and divide transversely in the centre, each half becoming a perfect new cell; those of mature threads have the endochrome in a dense mass, which eventually becomes a sporidium in their centre.

This is a summer annual, and grows near high-water mark in many localities on the British coast. It does not adhere very closely to paper when dried.

Hormotrichum collabens. The collapsing Hormotrichum.

Threads from three to six inches long, robust, those of the same tuft varying much in diameter, of soft, gelatinous texture, and bright verdigris-green colour; cells from once to once and a half the length of their diameter, filled with a dense mass of bright green granular endochrome, and having thick walls.

This species was formerly considered to be nearly allied to *Chætomorpha ærea;* but it always differed from that plant in external appearance, and is now found to possess distinct generic characters. It grows on floating wood, etc., in the sea, and is rare.

Hormotrichum bangioides. The Bangia-like Hormotrichum.

Threads from three to six inches long, very slender, smooth, soft, waved, attached at the base, growing in dense tufts, each thread of the same diameter throughout, but the threads of the same tuft varying much in size; cells about twice the length of their diameter, slightly constricted at the joints, each containing in its centre, surrounded by a broad transparent border, a dark-green mass of endochrome, that eventually becomes an oblong sporidium.

This species appears to be peculiar to Great Britain, and is not common even there. It grows on rocks near low-water mark, and the recorded habitats are on the south coast or in Ireland. In young specimens the endochrome nearly fills the cells; but the usual process of condensation takes place as the plant is developed, and eventually only a dark-green oblong spot is to be seen in the centre of each cell.

Hormotrichum Cutleriæ. Miss Cutler's Hormotrichum.

Threads delicately fine, curved, silky, growing in flowing tufts attached to pebbles, etc.; cells about the length of their diameter; those of young specimens filled with pale green, fluid endochrome, which is gradually condensed into a dark-green, globular sporidium that is eventually set free by the bursting of the mature cell.

This plant was formerly included in the genus *Lyngbya*. It was discovered by Miss Cutler, near the mouth of the river Otter, at Budleigh Salterton, and is named after her. It appears to grow on the sea-shore near the influx of fresh water, and where it is covered at high tide. The development of the endochrome varies, and the difference between a young and a mature thread is very marked.

Hormotrichum speciosum. The beautiful Hormotrichum.

Threads robust, soft, straight when young, becoming curled at maturity, from two to four inches long, growing in flowing or spreading tufts, attached to rocks and *Fuci;* cells not very distinctly divided, about half the length of their diameter, at first nearly filled with endochrome, which is condensed into an oblong sporidium, that eventually issues from the cell and leaves it colourless.

This species was also formerly a *Lyngbya*, and is nearly allied to *H. Carmichaelii*. It grows between the tide-marks during summer, and is annual. It is of a bright, glossy, yellow-green colour, and soft texture.

Hormotrichum Carmichaelii. Captain Carmichael's Hormotrichum.

Threads somewhat robust, several inches long, curled, tortuous, growing in densely interwoven layers of considerable extent on rocks and other substances; cells indistinctly divided, about a quarter the length of their diameter, at first filled with endochrome, which passes through the usual stages and issues from the cell at maturity, leaving it colourless.

This species closely resembles the last, and was also

included in the genus *Lyngbya*. It grows between the tide-marks during summer, and is annual. It has been found in many localities on the British coast. When growing, the threads are of a grass-green colour; but they become dull dark-green when dry. They adhere pretty well to paper.

Genus CXVI. RHIZOCLONIUM.

Frond a simple or imperfectly branched, decumbent, membranaceous thread, of uniform diameter throughout, and composed of a string of thin-walled, oblong cells filled with granular endochrome.—RHIZOCLONIUM, from the Greek *rhizoo*, to root, and *kloon*, a shoot.

The limits of this genus are not yet very exactly defined, for Kützing includes in it several species which Dr. Harvey refers to *Chætomorpha*, and Dr. Gray has added to it what was formerly *Lyngbya flacca*. So far as the British species are concerned, the characters to be observed are the short root-like branches, and the non-gelatinous substance of the threads.

Rhizoclonium riparium. The shore Rhizoclonium.

Threads slender, soft, long, angularly bent, spurred with short, root-like branches, densely woven into flat layers of considerable extent; cells about twice the length of their diameter, filled with pale green endochrome.

This plant grows during summer on rocks, which are covered with sand, usually near high-tide mark, and is annual. It has been found in several localities round our coast, but is far from common.

Rhizoclonium Casparyi. Dr. Caspary's Rhizoclonium.

Threads extremely slender, very soft, long, curved, spurred at distant intervals with short, root-like branches, woven into a thin web of considerable extent; cells varying in length, in the same layer, and even in a single thread, from once to six times the length of their diameter, filled with bright green granular endochrome, the grains of which are of unequal size.

This plant was first found by Dr. Caspary, at Falmouth and Penzance. It is altogether more slender and delicate than the normal state of *R. riparium*; but it is difficult to find definite characters whereon to base its claim to be considered a distinct species.

Rhizoclonium flaccum. The soft Rhizoclonium.

Threads about an inch long, very slender, silky, straight or slightly curved, simple or set with a few awl-shaped, rootlike branches, growing in patches on sea-weeds or floating wood, attached at the base; cells rather shorter than their diameter, at first nearly filled with grass-green endochrome, which is quickly condensed into a lens-shaped sporidium in the middle of the cell, where it appears surrounded by a broad transparent border.

Kützing has placed this species in the genus *Hormotrichum*, and Dr. Harvey in *Lyngbya*; but the latter author has at the same time expressed his approval of the position assigned to it by the former. In these circumstances I have hesitatated to deviate from the course thus prescribed. But after careful consideration I am of opinion that the root-like branches and the non-gelatinous substance of the threads are characters sufficient to justify the removal made by Dr. Gray in his 'Handbook of Water-weeds,' and which I have adopted. This plant is a summer annual, and may be found on many parts of our coast.

Order XXIV. OSCILLATORIACEÆ.

Green (rarely Olive-brown, Blue, or Purple), marine or freshwater Algæ, composed of simple or slightly-branched threads, each of which consists of an annular, medullary chord, of very short cells, enclosed in a one-celled, membranous sheath.

Only two or three genera of the plants belonging to this Order are marine. The remainder inhabit water of all degrees of quality and temperature, or grow in places which are merely damp. Among them are to be found the tenants of the Geysers of Iceland, and of the vats of poisonous chemicals, etc., to which I have already referred in the second chapter of this work. Certain kinds exhibit peculiar motions, similar to those of the lower forms of animal life, and are capable of progression and retrogression, and of ascent and descent in water. Speaking of the *Oscillatoriæ*, Dr. Harvey writes, "These movements are of three kinds: first there is the oscillating movement; one end of the thread remaining nearly at rest, while the other sways from side to side, sometimes describing nearly a quarter of a circle in a single swing. Secondly, the tip of the filaments has a minute movement, bending from side to side like the head of a worm; and, thirdly, there is an onward movement, probably the result of the two former. It is this latter which causes the filaments to radiate, and spread out from the edge of the stratum. If a minute portion of a living *Oscillatoria* be placed in water under a moderately high magnifying power, all these movements can be seen without trouble." These facts relate rather to the freshwater than to the marine species with which alone I am at present concerned, but they are so interesting in themselves that I could not refrain from recording them. I may add, that some species of *Oscillatoria* is to be found floating like a black or green scum on nearly every pool of stagnant water.

Genus CXVII. LYNGBYA.

Frond a simple, free, flexible, stationary, non-mucilaginous thread, composed of a continuous, cylindrical, membranaceous tube, filled with green or purple endochrome, which is densely annulated, and from which lens-shaped sporidia are developed.—LYNGBYA, in honour of H. C. Lyngbye, a Danish algologist.

Some of the species belonging to this genus are marine, and others grow in brackish or in fresh water. They are distinguished from the *Calothrices* by the greater length of their threads, and by their general habit, and from the *Oscillatoriæ* by the absence of a gelatinous matrix and of oscillating movement.

Lyngbya majuscula. The large Lyngbya.

Threads robust, tough, hair-like, twisted, growing in large tufts, which are densely interwoven in the centre, at first attached to the substance on which they grow, then floating; tube thick, forming a wide, transparent border to the dark-coloured endochrome.

This species grows on mud or sand, among rocks,

near low-water mark, and at greater depths. It is annual, and may be found during summer and autumn in many localities round our coast. The colour of the tufts is a very dark green, almost black.

Lyngbya ferruginea. The dark-blue Lyngbya.

Threads slender, soft, curved, growing on mud, in thin layers of considerable extent, or floating on the surface of stagnant pools of salt or brackish water; tube very thin, containing strongly annulated, bluish-green endochrome, with empty spaces at intervals.

This species is much smaller, and less conspicuous than *L. majuscula*, and in consequence is, no doubt, frequently overlooked. It grows on mud, in pools of brackish water, and is probably abundant in many localities, although but few habitats are recorded.

Genus CXVIII. **CALOTHRIX.**

Frond a separate, free, somewhat rigid, erect, stationary, non-mucilaginous thread, composed of a continuous tube, filled with densely annulated, green endochrome, from which lens-shaped sporidia are developed.—CALOTHRIX, from the Greek *kalos*, beautiful, and *thrix*, a hair.

This genus contains a large number of species, some of which are found in salt water, and some in fresh. They are all minute, and grow in tufts or patches on rocks, mud, or sea-weeds.

Calothrix Confervicola. The Conferva Calothrix.

Threads about one-tenth of an inch long, rigid, straight or slightly curved, blunt, opake, of a verdigris-green colour, growing in star-like tufts on the branches of filiform algæ; tube rather thick, containing separate masses of dull, bluish-green endochrome.

This very minute parasite is widely distributed, both on our own shores and on those of other countries.

Calothrix luteola. The yellowish Calothrix.

Threads about one-tenth of an inch long, flexible, blunt, transparent, of a pale yellowish colour, growing singly or in small tufts on filiform algæ; tube more or less filled with faintly annulated, pale green endochrome.

This species is very minute, and in consequence very difficult to collect. When found, it must be examined under a microscope of high power, or its structure and characters will not be properly revealed. The recorded observations of different authors do not agree; but the discrepancy has probably arisen from some of the specimens examined being young and others mature, or in a state of incipient decay. In the former, the threads, being full of endochrome, would be opake, while in the latter, the sporidia having been discharged, they would be empty and transparent, and this is the point that is in dispute.

Calothrix scopulorum. The rock Calothrix.

Threads about one-tenth of an inch high, simple, bent or curled, awl-shaped, tapered to a point, interwoven, growing in closely-packed, velvety patches, on rocks; tube filled with dull, yellowish-green endochrome, which is more or less distinctly annulated.

This species grows near high-water mark, and coats

the surface of all kinds of rocks with dull green, slippery layers of its tiny fronds. It is common in most localities, but too insignificant to attract attention. Its characters are of course microscopic.

Calothrix pannosa. The ragged Calothrix.

Threads from a quarter to half an inch long, rigid, strongly curled, blunt, of the same diameter throughout, closely woven into thin laminæ, which are arranged in a compact layer, the surface of which is sometimes irregular and ragged, sometimes divided into round or angular spaces, like a honeycomb; tube filled with strongly annulated, dark green endochrome.

This plant grows in pools near high-water mark, and is perennial. The two forms described above appear to depend on the position occupied by the specimens,—those which are parasitic on *Corallinæ*, etc., being irregular and ragged on the surface, while those which grow on rocks are honeycombed.

Calothrix semiplena. The variegated Calothrix.

Threads from half an inch to an inch or more long, very slender, tough, bent, blunt at the tip, closely woven into flat tufts, which are broad at the base, gradually tapered to a point at the top, stand erect, and are either loosely bundled together or piled one above another; tube partially filled with dense, verdigris-green endochrome, a long space of empty transparent tube alternating with a similar length of endochrome.

This species grows on *Corallinæ* and algæ in rock-pools, near high-water mark. It may be distinguished by its comparatively large size and variegated threads.

Genus CXIX. OSCILLATORIA.

Fronds simple, rigid, needle-like threads developed in and radiating from a gelatinous matrix, vividly oscillating; tube continuous, filled with densely annulated, green endochrome.—OSCILLATORIA, from the Latin *oscillum*, a swinging motion.

Most of the species belonging to this genus grow in fresh water; but some few are marine. Their most remarkable character is the oscillating, pendulum-like motion of the threads; this is more distinctly visible in some species than in others, and is alway greater in warm weather than in cold.

Oscillatoria litoralis. The shore Oscillatoria.

Threads thick, variously curved, growing in a bright verdigris-green layer; endochrome distinctly marked with closely set striæ, "divided at uncertain intervals into portions, which probably break off eventually and become new filaments."

This species was first found by Captain Carmichael, in muddy pools near the seashore, at Appin. Kützing considers it to be identical with *Lyngbya crispa* of Agardh.

Oscillatoria spiralis. The spiral Oscillatoria.

Threads slender, very short, much twisted, or spiral, closely interwoven, growing in dark or bluish-green, more or less leathery layers of indefinite dimensions; endochrome marked with somewhat distant, indistinct striæ.

This species has been found in Scotland, and in the south of England, on rocks, on submerged wood, and on dry bare earth. Kützing identifies it with his *Spirulina tenuissima*.

Oscillatoria nigro-viridis. The dark-green Oscillatoria.

Threads slender, rigid, blunt and curved at the tip, of a pale green colour, growing in thin, dark olive-green layers at first on the mud, then floating; endochrome marked with indistinct striæ, at intervals of about half the length of the diameter of the threads, not visibly granular.

Oscillatoria subuliformis. The awl-shaped Oscillatoria.

Threads very slender, awl-shaped, attenuated, acute and strongly curved at the tip, of a bright green colour, growing in thin, dark blue, almost black layers at first on the mud, then floating; endochrome marked with indistinct striæ at intervals of about three-fourths the length of the diameter of the threads, not visibly granular.

Oscillatoria insignis. The remarkable Oscillatoria.

Threads robust, somewhat brittle, rounded and adorned with minute, motionless cilia at the tip, of a brown colour, growing in thin dark-brown layers on decaying vegetable matter; endochrome marked at short intervals with distinct striæ, visibly granular.

The three last species are described in the 'Phycologia Britannica,' from specimens collected by Mr. Thwaites in brackish ditches, at Shirehampton, near Bristol. I cannot find them in Kützing's 'Species Algarum,' and I have no more recent work at hand to which I can refer. I am not myself acquainted with them.

Genus CXX. SPIRULINA.

Fronds simple, rigid, spirally twisted, vividly oscillating threads, developed in a thin gelatinous layer; tube continuous; endochrome green, more or less distinctly annulated. —SPIRULINA, from the Latin *spira*, a twist.

Spirulina tenuissima. The very slender Spirulina.

" Stratum very lubricous, æruginous, sub-radiant; filaments densely spiral, very slender, parallel, flexuous."— *Phyc. Brit.*

This species grows on decaying sea-weeds, or wood in brackish water.

Genus CXXI. TOLYPOTHRIX.

" Filaments of nearly equal diameter, tufted, tenacious ; branches free, continuous, with main filaments annulated at the base; cells indistinct, rarely moniliform."—TOLYPOTHRIX, from the Greek *tolupe*, a clew, or ball wound up, and *thrix*, hair.

I have taken the description of this genus from Dr. Gray's Handbook. The only British species that is marine was formerly *Calothrix fasciculata.*

Tolypothrix fasciculata. The fascicled Tolypothrix.

Threads straight, awl-shaped, tapered, at first simple,

then furnished with closely-pressed pseudo-branches, growing in dark-green, velvety layers of indefinite extent; endochrome dark-green, distinctly marked with close striæ.

Genus CXXII. ARTHRONEMA.

"Filaments simple, of equal diameter, united together longitudinally." (*Dr. J. E. Gray.*)—ARTHRONEMA, from the Greek *arthron*, a joint, and *nema*, a thread.

The marine species of this genus were also formerly *Calothrices*.

Arthronema hydnoides. Hydnum-like Arthronema.

Threads growing in flat, dark olive-green patches, of considerable extent, elongated, slightly curved, blunt, interwoven below, their points massed together in erect toothlike bundles; endochrome dark-green, distinctly striate at short intervals, surrounded by a broad, transparent, yellowish border.

This species grows on clayey seashores, near high-water mark, and is not uncommon. The threads are spuriously branched, or rather attached one to another, generally a small thread to a large one, by their sides near the tip in a longitudinal series.

Arthronema cæspitula. The cushioned Arthronema.

Threads growing in close, convex, blackish-green tufts, curved, twisted, soft, blunt, attached to each other longitudinally; endochrome distinctly striate at short intervals, surrounded by a narrow, transparent border.

Genus CXXIII. MICROCOLEUS.

Frond composed of minute, rigid, straight, ringed threads, enclosed many together in simple or branched, transparent sheaths, which are either open or closed at the upper end. —MICROCOLEUS, from the Greek *mikros*, small, and *koleos*, a sheath.

Microcoleus anguiformis. The snake-shaped Microcoleus.

Sheaths simple, decumbent, snake-like, broad and open at the tip, tapering gradually to the base, interwoven and twisted into irregular groups; threads short, slender, straight, striated at distant intervals.

This minute plant grows in dark-green layers on mud in brackish pools.

Genus CXXIV. SCHIZOTHRIX.

" Filaments involved in a thick lamellar sheath, rigid, curled, thickened at the base, at length longitudinally divided; spermatia lateral." (*Kützing.*)—SCHIZOTHRIX, from the Greek *schizo*, to divide, and *thrix*, a hair.

Schizothrix Cresswellii. The Rev. R. Cresswell's Schizothrix.

Threads exceedingly slender, once or twice forked, apparently by the division of the original thread, massed together in rope-like, branching bundles, which form roundish or oval patches, about two inches in diameter, and half an inch thick.

This plant was discovered on sandstone rock, near high-tide mark, under a drip of fresh water, at Sidmouth. It grows during winter, and is annual.

Genus CXXV. SCHIZOSIPHON.

Frond a globose or lobed gelatinous crust, composed of closely packed, ringed, radiating, sheathed threads, each of which springs from a transparent cell; sheath gelatinously membranous, cleft into numerous longitudinal fibres.—SCHIZOSIPHON, from the Greek *schizo*, to divide, and *siphon*, a tube.

Schizosiphon Warreniæ. Miss Warren's Schizosiphon.

"Fastigiately branched, the lowest cell of the branches wider, hemispherical, lateral; sheaths dark-coloured, the fibres often spiral; apices of the branches much attenuated."—*Caspary*.

Genus CXXVI. RIVULARIA.

Frond irregularly globose, fleshy, composed of continuous, radiating threads, which are annulated within, and have a transparent, spherical cell at the base.—RIVULARIA, from the Latin *rivulus*, a small brook.

Rivularia plicata. The wrinkled Rivularia.

Fronds of a dark green colour and irregular shape, gelatinous, from one-tenth of an inch to half an inch in diameter, wrinkled, hollow, often ruptured, growing in confluent groups; threads tapering gradually to a fine point, arranged in dichotomous series.

Rivularia atra. The black Rivularia.

Fronds of a glossy black colour, minute, globose, or hemispherical, smooth, firm, growing singly; threads awl-shaped, closely packed, of a dark green colour.

This species is very common, and grows at all seasons on rocks and sea-weeds, between the tide-marks.

Rivularia nitida. The shining Rivularia.

Fronds of a shining, bluish-green colour and irregular shape, from half an inch to an inch or more in diameter, gelatinously leathery, wrinkled, at first flat and solid, then more or less globose and hollow, growing in confluent groups; threads tapering to a very fine point, arranged in a somewhat dichotomous manner; those in the centre of the frond distant from each other, those at the surface closely packed.

This is the largest and most conspicuous marine species of British *Rivulariæ*. It grows on bare exposed rocks between the tide-marks, during summer and autumn, and is annual. It is common on the south coast, in Ireland, and in the Channel Islands.

Genus CXXVII. **ACTINOTHRIX.**

"Frond stellate; filaments elongate, subcylindrical, rather flaccid, radiating from a free central mass; when young this mass is large and spherical, and the fibres are short and conical, giving the whole plant the appearance of a *Calthrops*. As the algæ grow the filaments gradually elongate, become more cylindrical, that is, less conical and tapering, and the central mass decreases in size until, in the perfect plant, the long filaments seem to spring from a central dot; the endochrome is annulated, the rings numerous and very narrow, looking as a series of coins arranged closely side by side would appear if placed within a glass tube." (*Dr. J. E. Gray*, 'Journal of Botany,' n. xxiv.)—ACTINOTHRIX, from the Greek *aktis*, a ray, and *thrix*, hair.

Actinothrix Stokesiana. Mrs. Stokes's Actinothrix.

" Filaments bright green, about a quarter of an inch long, growing nineteen or twenty from one centre; endochrome surrounded by a transparent margin."

The specimens on which this genus and species are founded were collected by Mrs. Stokes, floating among *Cladophoræ*, in Dingle Bay, Ireland. Others have since been found on *Lyngbya majuscula*, at West Cowes.

ORDER XXV. NOSTOCHINEÆ.

Green, freshwater or rarely marine Algæ, composed of necklace-like threads, lying in a gelatinous matrix. Threads formed of globose cells, here and there interrupted by a single cell (heterocyst) of a different character. Propagation by zoospores.

Genus CXXVIII. **MONORMIA.**

Frond a branched mass of loose-textured gelatine, containing a spiral, necklace-like string of spherical, coloured, ordinary cells, with here and there a connecting-cell of larger size. Zoospores developed from the ordinary cells.—MONORMIA, from the Greek *monos*, one, and *hormos*, a necklace.

Monormia intricata. The intricate Monormia.

Fronds "forming small, roundish, gelatinous masses, floating among *Lemna* in fresh water, but probably within the influence of the tide; and also amongst *Enteromorpha intestinalis*, and even within its frond, in brackish water."

Genus CXXIX. **SPHÆROZYGA.**

Frond a gelatinous layer, or skin containing simple, necklace-like, waved or curved threads, composed of a series of ordinary cells, with connecting-cells of a different kind, usu-

ally spherical, of larger size and ciliated, at intervals. Fructification, large zoospores developed in the ordinary cells.—
SPHÆROZYGA, from the Greek *sphaira*, a sphere, and *zygos*, a yoke.

Sphærozyga Carmichaelii. Captain Carmichael's Sphærozyga.

Frond a thin, gelatinous skin, of a vivid green colour, containing minute, straight or slightly curved, necklace-like threads, which taper towards both ends. Zoospores large, oblong, twice or thrice the length of their diameter.

This plant grows on decaying algæ, in the sea, or in brackish ditches.

Sphærozyga Thwaitesii. Thwaites's Sphærozyga.

Frond very gelatinous, of a dark green, sometimes almost black colour, containing pale green, curved, entangled threads; connecting-cells large, somewhat oblong, ciliated, of a pale colour, sometimes terminal; ordinary cells somewhat flattened. Zoospores dark brown.

Sphærozyga Broomei. Broome's Sphærozyga.

Threads very slender; connecting-cells smooth, somewhat square, rather longer than wide. Zoospores numerous, elliptical, about twice the length of their diameter, and only a little wider than the ordinary cells.

This species was discovered by Mr. G. E. Broome, on dead leaves of *Myriophyllum*, etc., in a brackish ditch at Shirehampton, near Bristol.

Sphærozyga Berkeleyana. The Rev. M. J. Berkeley's Sphærozyga.

Threads when young enclosed, one or several together,

in a defined, transparent sheath; connecting-cells spheroidal, slightly flattened. Zoospores large, oblong, about once and a half as long as broad, and twice the diameter of the ordinary cells, generally formed in pairs on each side of the connecting-cell, dark brown when ripe.

This species was found growing on *Cladophora fracta*, in a brackish ditch, at Shirehampton.

Genus CXXX. SPERMOSIRA.

Threads lying in dark-green, somewhat gelatinous layers, simple, cylindrical, composed of a series of lens-shaped ordinary cells, with larger, compressed, connecting-cells at intervals, each thread enclosed in a transparent, membranous tube. Zoospores formed from the ordinary cells.—SPERMOSIRA, from the Greek *sperma*, a seed, and *seira*, a string.

The tube which envelopes the thread is the character by which this genus is distinguished from *Sphærozyga*.

Spermosira litorea. The shore Spermosira.

Frond a dark green, fleecy layer, composed of nearly straight, robust threads; ordinary cells short, compressed, of a blue-green colour; connecting-cells pale-red. Zoospores elliptical, dark brown.

This species grows on floating plants in muddy, brackish ditches.

Spermosira Harveyana. Dr. Harvey's Spermosira.

Frond a layer of much curved, slender threads; ordinary cells nearly the length of their diameter; connecting-cells somewhat square, rather longer than their diameter, which

EXPLANATION OF SCIENTIFIC TERMS

USED IN THIS WORK.

Abnormal, out of natural order.
Aculeate, prickly.
Acuminate, pointed.
Acute, sharp.
Æruginous, of the colour of verdigris.
Algological, relating to water-weeds.
Alternate, branches, etc., growing on either side the stem, at equal distances, not opposite.
Anastomosing, uniting of vessels, veins, or nerves.
Anceps, two-edged.
Annual, lasting one year only.
Annular, } ringed.
Annulate,
Anther, the part of a flower containing fertilizing matter.
Antheridium, an organ supposed to contain fertilizing matter.
Arbuscula, a shrub.
Apex, the top.
Appressed, pressed close to a stem or branch.
Articulate, jointed.
Articulation, strictly the point of junction of two parts of a plant, but frequently applied to the part of a stem intervening between two joints.
Attenuate, tapering gradually to a point.

EXPLANATION OF SCIENTIFIC TERMS. 299

Axial, appertaining to the axis.
Axil, the angle formed by the union of the stem and branch, or of two branches.
Axillary, placed in the axils.
Axis, the central column of a cylindrical frond.
Basal, at or near the base.
Biennial, lasting two years.
Bifid, twice cut.
Bipinnate, furnished with primary and secondary branchlets.
Branchlet, a small or secondary branch.
Byssoid, like the fibre of cotton, silk, or flax.
Calcareous, partially formed of chalk or lime.
Canaliculate, channelled, grooved.
Capsule, a spore-vessel.
Cartilage, gristle.
Cartilaginous, gristly.
Cellular, composed of cells.
Cellule, a small cell.
Ciliate, fringed with hair-like processes.
Cilium, a hair-like process or fibril—literally an eyelash.
Claviform, club-shaped.
Compressed, slightly flattened.
Concentric, arranged round a common centre.
Conceptacle, an organ containing spores.
Confervoid, conferva-like.
Conical, shaped like a cone.
Convergent, tending towards the same point.
Constricted, drawn together, narrowed.
Convex, rising in a circular form.
Cordate, heart-shaped.
Coriaceous, leathery.
Corpuscle, a small body.
Cortical, belonging to the back.
Cruciate, in the form of a cross.
Crustaceous, crust-like, and coated with chalk or lime.

Cuticle, a fine membrane, usually transparent, which covers the surface of a plant.
Cylindrical, long and round, like a rod.
Deciduous, falling off.
Decumbent, leaning downwards.
Decurrent, attached to the stem without a stalk, and extending along it.
Denuded, laid bare.
Depauperated, impoverished.
Dichotomous, branched by repeated forkings.
Digitate, finger-shaped.
Diœcious, having male and female flowers on distinct plants.
Dissepiment, an internal partition.
Distichous, ranged in two opposite rows.
Divaricate, growing in many different directions.
Divergent, very open and spreading.
Echinate, prickly, spined.
Elliptic, about twice as long as broad, with rounded ends equal to each other.
Elongate, long and narrow.
Endochrome, coloured matter contained in the cells of sea-weeds, etc.
Epidermis, the outer coating of cellular tissue.
Esculent, eatable.
Exotic, foreign, not native.
Fasciculate, growing together in a tuft.
Fastigiate, level-topped—when the branches all point upwards, as in the Lombardy poplar.
Favella, a globose cluster of spores.
Favellidium, a cluster of two or more favellæ.
Fibril, a very delicate fibre.
Fibrillose, covered with fibres.
Fibrous, composed of fibres.
Filamentous, composed of long, simple or branched threads.
Filiform, formed like a thread.

Flaccid, soft, limp.
Flexuose, irregularly bent.
Fluviatile, growing in rivers.
Forcipate, pincer-shaped.
Frond—of a sea-weed, the whole plant, except the root and fructification.
Fructification—of sea-weeds, etc., the organs of propagation.
Gelatino-membranaceous, gelatinously membranaceous.
Gelatinous, like a jelly.
Gemma, a bud.
Gemmule, a small bud that falls off and becomes a new plant.
Generic, appertaining to a genus.
Genus, a group of species having a common character in their fructification.
Glaucous, having a grey tint on the surface.
Glutinous, sticky.
Granule, a small grain.
Granular, } composed of granules.
Granulate,
Homogeneous, of uniform substance or structure.
Hyaline, transparent and colourless.
Imbricate, overlapping.
Inarticulate, not jointed.
Incurved, bent inwards.
Indigenous, native.
Inflated, swollen.
Infusoria, microscopic animalcules.
Interlaced, mixed together.
Internode, part of a stem or branch between two nodes.
Interruptedly-pinnate, leaflets alternately large and small.
Interstices, a small intervening space.
Involucral, having an involucre.
Involucres, the branchlets which surround spore-clusters.
Laciniæ, small, irregular divisions of a flat frond.

Laciniate, jagged.
Lamella, a thin plate.
Lamina, the surface of a frond.
Lanceolate, lance-shaped.
Lateral, on the side.
Lax, loose, not dense.
Leaflet, a small leaf.
Leathery, tough, flexible, rather thick.
Ligulate, strap-shaped.
Linear, long and narrow, with parallel sides.
Lobe, a division of a flat frond.
Lobule, a lesser division of a flat frond.
Longitudinal, lengthwise.
Lubricous, smooth, slippery.
Matrix, the substance in which anything is developed or deposited.
Medullary, pith-like.
Membranous,
Membranaceous, } soft, supple, rather thin.
Midrib, the vein-like thickening running down the centre of a frond.
Moniliform, necklace-like.
Monœcious, having male and female flowers on the same plant.
Mucilaginous, slimy.
Mucilage,
Mucus, } a gelatinous liquid or slime.
Multifid, many times cut or divided.
Multipartite, many-parted, divided to the base.
Nemathecia, external wart-like excrescences.
Nerves, fibres visible on the external surface of a frond.
Node, the place where the joints or articulations of a stem meet.
Nodose, knotted.
Normal, ordinary condition.

EXPLANATION OF SCIENTIFIC TERMS. 303

Nucleoli, a small cluster of spores forming part of a nucleus.
Nucleus, a mass of spores sometimes composed of several small clusters.
Obconical, inversely conical.
Oblong, longer than broad.
Obovate, about twice as long as broad, the upper end larger than the lower.
Obtuse, rounded at the top.
Orbicular, circular, spherical.
Orifice, an opening.
Oscillate, to move to and fro like a pendulum.
Ovate, egg-shaped, about twice as long as broad, the lower end larger than the upper.
Palmate, hand-shaped.
Paranemata, filaments that accompany spores.
Parasitic, growing upon other plants, and deriving nourishment from them. This word is improperly applied to sea-weeds, as they do not derive nourishment through their roots. They should be said to be epiphytic.
Patent, spreading.
Pectinate, set with comb-like branches or branchlets.
Pedicellate, stalked.
Pellucid, transparent.
Pendulous, drooping, hanging.
Perennial, lasting more than two years.
Perforate, pierced with small holes.
Pericarp, a spore vessel or its walls.
Periderm, a membrane surrounding a spore cluster.
Periphery, the outer layer of cells of a cylindrical frond.
Peripheric, belonging to the periphery.
Perispore, a membrane surrounding a spore.
Pinna, a primary branchlet or leaflet.
Pinnate, having only primary branchlets or pinnæ springing from either side.

Pinnatifid, divided into lobes from the margin nearly to the midrib.

Pinnule, a secondary branchlet or leaflet.

Pistil, the body in the centre of a flower which contains the ovary, styles, and stigma.

Placenta, that part of the capsule to which the spores or spore-threads are attached.

Plumulate, } feathered.
Plumose,

Plumule, a feather-like branchlet.

Pollen, the fertilizing powder contained in the cells of flowering plants.

Polysiphonous, many-tubed.

Pore, a minute opening in the cuticle.

Primary, of the first series.

Process, a prominence, protrusion, or small lobe.

Procumbent, spread on the ground, but not striking root.

Produced, lengthened.

Proliferous, producing leaflets or shoots from the tip or margin.

Punctate, dotted.

Pyramidal, formed like a pyramid.

Pyriform, pear-shaped.

Quadrifarious, arranged in four or more rows.

Racemose, having several conceptacles along a branchlet.

Ramellus, the ultimate division of a branch.

Ramus, a principal branch.

Ramulus, a secondary branch.

Receptacle, the vessel which contains the spores in the Olive series.

Rectangular, right-angled.

Recurved, } bent downwards.
Reflexed,

Reticulate, having a net-like structure or marking.

Rigid, stiff.

Saccharine, sugary.
Saline, salt.
Secondary, the first division of a primary branch.
Secund, arranged on one side only of a stem.
Segment, a division of a flat frond.
Serrate, saw-edged.
Sessile, without a stalk.
Seta, a bristle.
Setaceous, bristly, of the diameter of a bristle.
Sinuate, more or less deeply cut.
Sinuous, with a waved margin.
Siphon, a tube.
Sorus, a group of spores or tetraspores.
Specific, appertaining to species.
Spherical, concave or globular.
Spheroidal, like an oblong sphere.
Sporaceous nucleus, a spore-like cluster.
Sporangium, a case containing a spore or spores.
Spore, the equivalent in Cryptogamic plants of a seed in flowering plants.
Sporidium, a pseudo-spore.
Sporiferous nucleus, a cluster of spores.
Sporophyllum, a leaflet bearing spores or tetraspores.
Sporule, a small spore.
Squarrose, spreading at right angles.
Stichidium, a pod-like vessel containing tetraspores.
Stratum, a layer.
Stria, a line, flute, or stripe.
Striate, striped or fluted.
Subcordate, somewhat heart-shaped.
Subdichotomous, somewhat forkedly divided.
Subglobular, somewhat globular.
Submoniliform, slightly necklace-like.
Subsimple, very slightly branched.
Subsolitary, nearly solitary.

Subulate, awl-shaped.
Succulent, juicy.
Tenacious, tough.
Tetraspore, a four-parted spore, a name usually applied to one of the kinds of fructification of the Red sea-weeds.
Tomentose, cloth-like, nappy.
Tortuous, many times bent, twisted.
Transverse, crosswise.
Trichotomous, repeatedly divided in threes.
Tripartite, divided into three.
Tripinnate, when the secondary branchlets or pinnules of a frond are branched.
Truncate, not pointed, ending abruptly.
Tuber, a fleshy, longish or roundish root.
Tubercle, a wart or knot containing spores or tetraspores.
Tuberculate, covered with knots or warts.
Umbelliferous, bearing branchlets, springing from a common centre and forming round flat heads.
Vascular, veined.
Verrucose, warty.
Verticillate, whorled, placed in circles round a stem.
Vibratile, inclined to vibrate.
Villous, covered with numerous soft hairs.
Virgate, rod-like.
Whorls, branchlets arranged in a regular circumference round a stem.
Zigzag, angularly bent from side to side.
Zonate, divided or marked by transverse or circular lines or bands.
Zoospores, (from *zoos*, living, and *sporos*, a seed) the fructification of certain sea-weeds of the Green series, in which a ciliary motion gives the appearance of life.

INDEX.

[The names of Genera are printed in large type, of Species in ordinary type, and Synonyms in italics.]

	Page		Page
acanthonotum	208	bacciferum	37
acicularis	183	Balliana	267
ACTINOCOCCUS	155	BANGIA	249
ACTINOTHRIX	293	bangioides	278
aculeata	48	barbata (CYSTOSEIRA)	39
adhærens (CODIUM)	244	,, (GRIFFITHSIA)	219
,, (CRUORIA)	155	barbatum	236
ærea	273	Berkeleyana	295
affine	229	Berkeleyi	69
agariciformis	132	bifida	170
AHNFELTIA	179	,, var. incrassata	170
ALARIA	51	,, var. *ciliata*	170, 171
alata	135	BONNEMAISONIA	118
albida	264	Bonnemaisonii	141
amphibium	243	Borreri	227
amphibius	80	BOSTRYCHIA	92
anceps	44	*botryocarpum*	203
anguiformis	291	*brachiatum*	225
angustissima	136	brachiatus	84
appendiculata	171	Brodiæi (CALLITHAMNION)	224
arbuscula (CALLITHAMNION)	223	,, (PHYLLOPHORA)	175
,, (DASYA)	114	,, var. simplex	175
arcta	268	,, (POLYSIPHONIA)	99
arenicola	275	Broomei	295
arenosa	275	Brownii	265
ARTHROCLADIA	49	BRYOPSIS	246
ARTHRONEMA	290	bulbosa	53
articulata	187	bursa	242
asparagoides	118	byssoides	111
ASPEROCOCCUS	64	byssoideum	231
atomaria	59	Cabreræ	50
atra	292	cæspitosa	120
ATRACTOPHORA	162	cæspitula	290
atro-rubescens	100	calcarea	130
attenuata (CROUANIA)	214	CALLIBLEPHARIS	141
,, (STRIARIA)	62	CALLITHAMNION	221, 222

BRITISH SEA-WEEDS.

	Page
CALLOPHYLLIS	181
CALOTHRIX	284, 289, 290
canaliculatus	44
capillaris	196
Carmichaelii (HOMOTRICHUM)	279
„ (SPHÆROZYGA)	293
CARPOMITRA	50
CASPARYI	281
CATENELLA	194
Cattlowiæ	117
ceramicola	251
CERAMIUM	200
ceranoides	42
CHÆTOMORPHA	272, 280
CHAMPIA	122
CHONDRIA	88
CHONDRUS	176, 186
CHORDA	54
CHORDARIA	66
Chrysymenia	187
CHYLOCLADIA	122, 123, 187
ciliaris	250
ciliata	145
ciliatum	209
cirrhosa	78
CIADOPHORA	259
CLADOSTEPHUS	75
clathrata	253
„ var. Linkiana	254
„ var. erecta	254
„ var. ramulosa	254
clavæformis	85
clavatum	75
claveilosa	188
coccinea (DASYA)	113
„ (DUDRESNAIA)	212
coccineum	171
CODIUM	242
collabens	277
collaris	58
complanata	94
compressa (ENTEROMORPHA)	253
„ (GRACILARIA)	147
compressus	64
Conferva	272, 276, 277
confervicola	284
confervoides	147
CORALLINA	125
corallina	217
CORDYLECLADIA	172

	Page
corneum	150
corniculata	128
cornucopiæ	252
coronopifolius	148
corymbosum	231
CORYNOSPORA	221
Cresswellii	291
crinitus	82
crispa	69
crispus	186
cristata	169
CROUANIA	213
cruciatum	234
cruenta	154
CRUORIA	154
curta	72
CUTLERIA	55
Cutleriæ	278
CYSTOCLONIUM	180
CYSTOSEIRA	38
Cystothrix	272
CYTOPHORA	276
DASYA	111
Dasyclonia	111
dasyphylla	89
„ var. squarrosa	89
Daviesii	238
decurrens	202
DELESSERIA	134
dentata	87
Deslongchampsii	204
DESMARESTIA	46
Devoniensis	218
diaphanum	203
dichotoma	60
„ var. intricata	60
DICTYOSIPHON	62
DICTYOTA	60
diffusa	262
digitata	52
„ var. stenophylla	52
distortus	82
divaricata (CHORDARIA)	67
„ (HELMINTHORA)	157
Dubyi (PEYSSONELIA)	153
„ (SCHIZYMENIA)	194
DUDRESNAIA	156, 212
DUMONTIA	196
echinatus	65
echionotum	207

INDEX. 309

	Page
ECTOCARPUS	79
edulis	193
ELACHISTA	70
elegans (BANGIA)	251
„ (PTILOTA)	211
elongata	107
elongella	107
ENTEROMORPHA	251
equisetifolius	215
erecta (CORDYLECLADIA)	172
erecta (ENTEROMORPHA)	254
ericoides	39
esculenta	51
EUTHORA	168
falcata	263
farinosa	132
fascia	54
fasciculata (MELOBESIA)	131
„ (TOLYPOTHRIX)	289
fasciculatum	227
fasciculatus	81
fastigiata (FURCELLARIA)	191
„ (POLYSIPHONIA)	110
fastigiatum	207
fenestratus	81
ferruginea	284
fibrata	104
fibrillosa	109
fibrosa	40
filamentosa	198
filicina (GRATELOUPIA)	192
„ (SPHACELARIA)	77
filiformis (DUMONTIA)	197
„ var. crispata	197
„ (MYRIOTRICHIA)	86
filum	55
flabelligerum	210
flaccida	71
flaccum	281
flagelliformis	66
flavescens	266
flexuosa	264
floccosum	233
floridulum	237
fœniculacea	40
fœniculaceus	62
formosa	104
fructa	267
fruticulosa	96
fucicola	71

	Page
FUCUS	41
FURCELLARIA	190
furcellata (POLYSIPHONIA)	103
„ (SCINAIA)	159
fusca	78
fusco-purpurea	250
Gattyæ	270
GELIDIUM	136, 149
GIGARTINA	183
Ginnania	159
glandulosa	199
glaucescens	263
GLOIOSIPHONIA	195
Gmelini	142
GRACILARIA	146, 172
gracilis	265
gracillimum(CALLITHAMNION)	230
„ (CERAMIUM)	205
granulata	39
granulosus	84
GRATELOUPIA	191
Grevillei	73
GRIFFITHSIA	215
Griffithsiæ	178
Griffithsiana (MESOGLOIA)	68
„ (POLYSIPHONIA)	105
„ (SEIROSPORA)	220
GYMNOGONGRUS	177, 179
HALIDRYS	37
HALISERIS	56
HALURUS	214
HALYMENIA	189
HAPALIDIUM	133
Harveyana	296
HELMINTHOCLADIA	158
HELMINTHORA	156
Hennedyi	155
HILDENBRANDTIA	153
Hilliæ	141
HIMANTHALIA	45
Hincksiæ	81
Hookeri	226
Hopkirkii	255
HORMOTRICHUM	276
Hutchinsiæ	262
hydnoides	290
Hypnea	180
hypnoides (ATRACTOPHORA)	163
„ (BRYOPSIS)	247
hypoglossum	137

	Page		Page
hystrix	297	Mackaii	44
implexa	275	Magdalenæ	266
insignis	288	majuscula	283
interrupta	174	mamillosa	185
interruptum	232	marina	245
intestinalis	252	MAUGERIA	164
intricata	294	melagonium	273
Iridæa	193	MELOBESIA	129
JANIA	127	membranacea	132
jubata	145	membranifolia	176
kaliformis	123	Mertensii	85
KALLYMENIA	182, 194	mesocarpum	238
laceratum	142	MESOGLOIA	67
laciniata	181	MICROCLADIA	199
laciniata	219	MICROCOLEUS	291
lactuca	257	microphylla	183
lætevirens	261	MONORMIA	294
LAMINARIA	51	multifida (CUTLERIA)	56
laminariæ	66	,, (WRANGELIA)	161
Landsburgii	82	multifidum	157
lanosa	269	multipartita	146
latifolia	63	MYRIONEMA	73
latissima	257	MYRIOTRICHIA	85
LAURENCIA	88, 119	NACCARIA	161, 163
LEATHESIA	68	*Nemaleon*	159
Lechlancherii	74	NEMALION	157
LEPTOCYSTEA	258	*Nicæensis*	168
lichenoides	132	nigrescens	99
ligulata (DESMARESTIA)	47	nigro-viridis	288
,, (HALYMENIA)	189	nitida	293
linearis	249	NITOPHYLLUM	139
Linkiana	254	*nodosum*	xxi
linum	274	nodosus	43
linza	256	Norvegicus	178
Lithocystis	xix, 133	nuda	267
litoralis (ECTOCARPUS)	83	obscura	1·1
,, (OSCILLATORIA)	287	obtusa	121
litorea (CYPTOPHORA)	276	ocellata	113
,, (SPERMOSIRA)	296	OCHLOCHÆTE	297
LITOSIPHON	65	ODONTHALIA	87
LOMENTARIA	123	officinalis	126
lomentaria	55	*Oligosiphonia*	98
longicruris	53	opuntia	195
longifructus	83	OSCILLATORIA	287
lorea	46	ovalis	125
luteola	285	PADINA	57
lycopodioides	90	palmata	167
LYNGBYA	279, 280, 281, 283	,, var. marginifera	167
Lyngbyæi	61	,, var. simplex	167
Macallana	262	,, var. Sarniensis	167

INDEX.

	Page
palmata, var. sobolifera	168
palmetta	168
palmettoides	177
pannosa	286
parasitica	101
parvula (CHAMPIA)	122
,, (*Chylocladia*)	122
,, (ZONARIA)	59
pavonia	57
pedicellatum	221
pedunculatus	50
pellita	155
pellucida	259
percursa	255
PETROCELIS	154
Peyssonelia	152
Phycoseris	256
phyllactidium	133
phyllitis	54
PHYLLOPHORA	175
pinastroides	94
pinnatifida	120
pinnatinervia	47
pistillata	184
plantaginea	63
plicata (AHNFELTIA)	179
,, (*Gymnocongrus*)	179
,, (RIVULARIA)	292
PLOCAMIUM	171
pluma	235
plumosa (BRYOPSIS)	246
,, (PTILOTA)	211
,, (SPHACELARIA)	77
plumula	234
POLYIDES	151
polymorpha	131
polypodioides	57
POLYSIPHONIA	97
polyspermum	228
PORPHYRA	248
PTILOTA	210
pulvinata (ELACHISTA)	72
,, (POLYSIPHONIA)	105
PUNCTARIA	63
punctatum	140
,, var. ocellatum	140
,, var. crispatum	140
,, var. Pollexfenii	140
,, var. fimbriatum	140
punctiforme	74

	Page
punicea	116
purpurascens	180
,, var. cirrhosa	180
purpurea	158
pusillus (ECTOCARPUS)	82
,, (LITOSIPHON)	65
pustulata	133
PYCNOPHYCUS	40
racemosa	79
radicans	78
RALFSIA	70
Ralfsii	255
ramulosa	254
rectangularis	261
reflexa	124
refracta	264
reniformis	182
repens	271
RHIZOCLONIUM	280
rhizodes	61
RHODOMELA	90
RHODOPHYLLIS	169
RHODYMENIA, 145, 166, 169, 170,	182
Richardsoni	106
riparium	280
RIVULARIA	292
rosea	188
roseum	232
rostratum	136
Rothii	237
rotundus	152
rubens (JANIA)	128
,, (PHYLLOPHORA)	176
rubra	153
rubrum	201
,, var. decurrens	202
,, var. pedicellatum	203
,, var. proliferum	203
,, var. secundatum	203
Rudolphiana	272
rupestris	260
,, var. distorta	261
ruscifolia	138
RYTIPHLŒA	93
saccharina	53
sanguinea	165
SARGASSUM	36
SCHIZOSIPHON	292
SCHIZOTHRIX	291

	Page		Page
SCHIZYMENIA	193	tenuissima (SPIRULINA)	289
SCINAIA	159	tenuissimum	205
scoparia	77	tetragonum	225
scopulorum	285	,, var. brachiatum	225
scorpioides	92	tetricum	225
scutulata	72	thuyoides	95
secundiflora	217	thuyoideum	230
SEIROSPORA	219	Thwaitesii	295
semiplena	286	TOLYPOTHRIX	289
sericea	212	tomentosum	213
serratus	43	tomentosus	81
setacea	216	tortuosa	274
siliculosus	80	tripinnatum	229
siliquosa	38	tuberculatus	41
simplicifilum	215	tuberiformis	69
sinuosa	134	Turneri (ASPEROCOCCUS)	64
sparsum	239	,, (CALLITHAMNION)	235
speciosum	279	ULVA	256
SPERMOSIRA	296	uncialis	270
SPHACELARIA	76	uncinatum	143
SPHÆROCOCCUS	148	urceolata	103
sphærophorus	84	*Vagabundia*	268
SPHÆROZYGA	294	variegata	244
spinulosa	106	VAUCHERIA	102
spiralis	287	velutina	73
SPIRULINA	288, 289	venusta	114
spongiosum	224	vermicularis	67
spongiosus	76	verrucata	133
SPOROCHNUS	49	verrucosa	70
SPYRIDIA	198	versicolor	144
squamata	127	*Vertebralia*	110
stellulata	71	verticillatus	75
STENOGRAMMA	173	vesiculosus	42
STILOPHORA	61	villosa	49
Stokesiana	294	violacea	108
strangulans	74	virescens	68
STRIARIA	62	virgatulum	239
strictum	206	viridis	48
subfusca	91	vulgare	36
submarina	245	vulgaris	248
subulifera	100	Warreniæ	292
subuliformis	288	Wiggii	162
sutoria	273	*Wormskioldia*	165
TAONIA	59	WRANGELIA	160
Teedii	185	Younganum	277
tenuissima (CHONDRIA)	89	ZONARIA	58
,, (PUNCTARIA)	63		

PLATE I.
(*Frontispiece.*)

FIG.   PAGE

1. Sargassum vulgare 36
    1 *a*, air-vessel.
    1 *b*, spore-receptacles.

2. Halidrys siliquosa 38
    2 *a*, section of receptacle, showing spores, *magnified*.

3. Pycnophycus tuberculatus 41
    3 *a*, section of receptacle, showing spores, *magnified*.

4. Leathesia tuberiformis 69
    4 *a*, filaments with spores.

PLATE II.

FIG.   PAGE

1. Fucus vesiculosus 42

    1 *a*, section of receptacle, showing spores, *magnified*.

2. Asperococcus Turneri 64

    2 *a*, part of frond, showing spore-clusters, *magnified*.

3. Laurencia cæspitosa 120

    3 *a*, tip of branch, showing tetraspores, *magnified*.

II.

Vincent Brooks imp

PLATE III.

FIG.   PAGE

1. Fucus nodosus 43
    1 *a*, segment of a spore-receptacle.
2. Fucus anceps 44
    2 *a*, pointed spore-receptacle.
    2 *b*, branchlet with antheridia.
3. Haliseris polypodioides 57
    3 *a*, portion of frond with sorus.
4. Sphacelaria filicina 77
    4 *a*, a pinna, *magnified*.

Vincent Brooks imp

PLATE IV.

FIG.   PAGE

1. Fucus canaliculatus 44

    1 *a*, part of a spore-receptacle.

    1 *b*, section of spore-receptacle.

2. Delesseria ruscifolia 138

    2 *a*, tip of leaf with spore-conceptacle, *magnified*.

3. Schizymenia edulis 193

    3 *a*, section of frond, showing spore-clusters, *magnified*.

W. Fitch. lith.

Vincent Brooks. Imp.

PLATE V.

FIG.   PAGE

1. Alaria esculenta 51

    1 *a*, section of part of a sorus, *magnified*.

2. Laminaria saccharina 53

    2 *a*, slice of frond, *magnified*.

W Fitch. lith

2a

Vincent Brooks. Imp

PLATE VI.

FIG.   PAGE

1. Laminaria Phyllitis 54

     1 *a*, slice of frond, *magnified*.

2. Lomentaria ovalis 125

     2 *a*, a branchlet with spore-conceptacles, *magnified*.

     2 *b*, a branchlet with tetraspores, *magnified*.

3. Porphyra vulgaris 248

     3 *a*, vertical section of frond, *magnified*.

W. Fitch, lith.　　　　　　　　　　　　Vincent Brooks, Imp.

PLATE VII.

FIG.   PAGE

1. Chorda lomentaria 55

    1 *a*, transverse section of part of frond, *magnified*.

2. Padina pavonia 57

    2 *a*, recurved margin, *magnified*.

    2 *b*, fringe, *magnified*.

    2 *c*, young sorus, *magnified*.

    2 *d*, old sorus, *magnified*.

3. Dictyota dichotoma 60

    3 *a*, sorus.

4. Elachista fucicola 71

    4 *a*, branched thread of tubercle, with spore, *magnified*.

5. Melobesia polymorpha 131

    5 *a*, portion of frond, showing spore-conceptacles, *magnified*.

PLATE VIII.

FIG.   PAGE

1. Odonthalia dentata 87

    1 *a*, branchlet with spore-conceptacles, *magnified*.

    1 *b*, branchlet with stichidia, *magnified*.

2. Nitophyllum punctatum 140

    2 *a*, sorus, *magnified*.

3. Catenella opuntia 195

    3 *a*, fronds, *magnified*.

    3 *b*, branch with spore-conceptacle, *magnified*.

PLATE IX.

FIG.   PAGE

1. Polysiphonia parasitica 101
    1 *a*, branchlet with spores, *magnified*.
    1 *b*, branchlet with tetraspores, *magnified*.
2. Gelidium corneum 150
    2 *a*, branchlet with spores, *magnified*.
    2 *b*, branchlet with tetraspores, *magnified*.
3. Halymenia ligulata 189
    3 *a*, section of frond with spores, *magnified*.
4. Griffithsia corallina 217
    4 *a*, part of branch with tetraspores, *magnified*.
    4 *b*, spore-clusters, *magnified*.

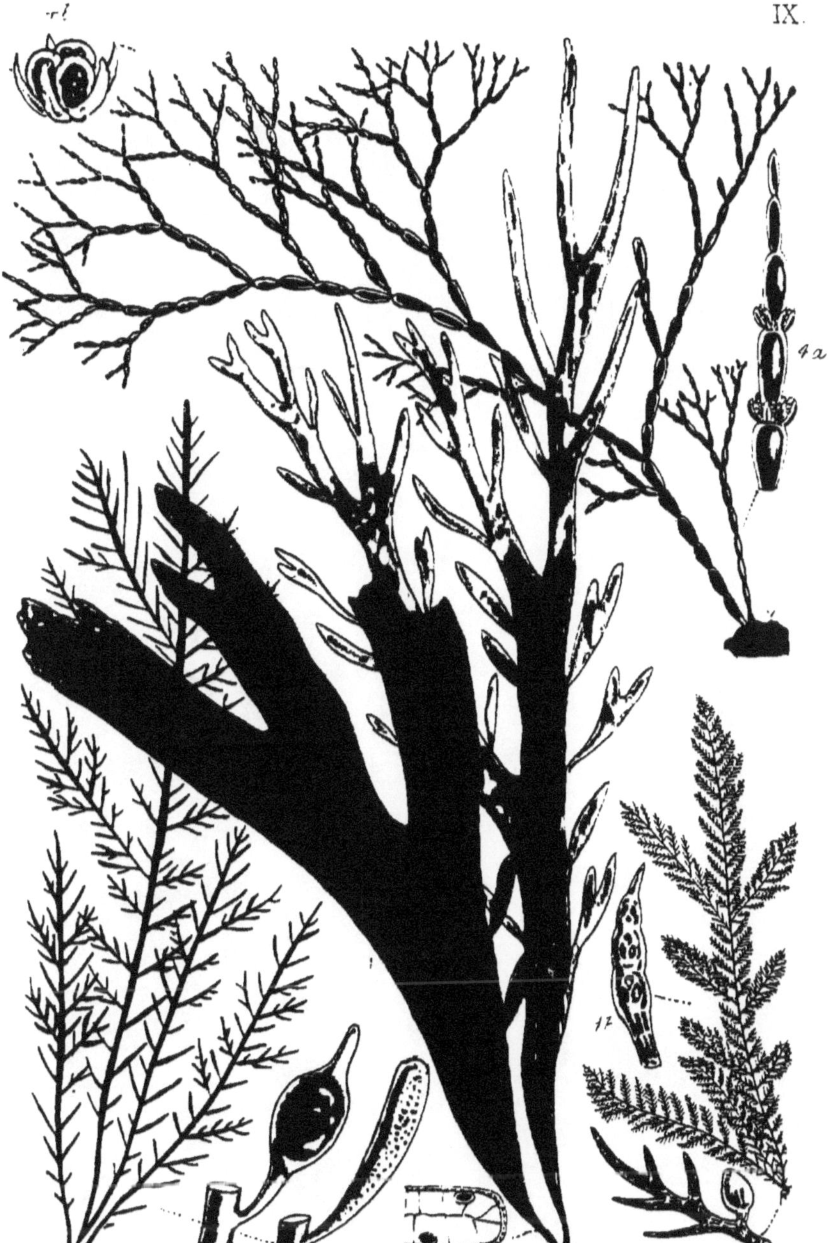

V. Fitch lith. Vincent Brooks Imp

PLATE X.

FIG.   PAGE

1. Bonnemaisonia asparagoides 118
    1 *a*, branch with spore-conceptacles, *magnified*.
2. Wrangelia multifida 161
    2 *a*, part of a branch, *magnified*.
    2 *b*, tetraspores, *magnified*.
3. Callophyllis laciniata 181
    3 *a*, spore-conceptacles, *magnified*.

W. Fitch lith. Vincent Brooks. Imp

PLATE XI.

FIG.   PAGE

1. Nitophyllum laceratum 142
    1 *a*, marginal processes with tetraspores, *magnified*.
2. Chylocladia articulata 187
    2 *a*, part of a branch with spores, *magnified*.
3. Gloiosiphonia capillaris 196
    3 *a*, branchlet with fructification, *magnified*.
4. Callithamnion plumula 234
    4 *a*, spore-clusters, *magnified*.

PLATE XII.

FIG.   PAGE

1. Sphærococcus coronopifolius 148

    1 *a*, branchlet with spores, *magnified*.

2. Maugeria sanguinea 165

    2 *a*, midrib with spore-leaflets.

3. Ptilota plumosa 211

    3 *a*, comb-like branchlet, *magnified*.

    3 *b*, spore-clusters, *magnified*.

PLATE XIII.

FIG.   PAGE

1. Desmarestia aculeata   48
    1 *a*, branchlet of a young frond, *magnified*.
2. Cladostephus verticillatus   75
    2 *a*, whorls of branchlets, *magnified*.
    2 *b*, branchlets, *magnified*.
    2 *c*, branchlets with spores, *magnified*.
3. Ulva latissima 257
    3 *a*, cellules of upper layer of frond, *magnified*.

N Fitch lith. Vincent Brooks, Imp

PLATE XIV.

FIG.   PAGE

1. Sporochnus pedunculatus 50

    1 *a*, a mature receptacle, *magnified*.

2. Codium bursa 242

    2 *a*, fibres of frond, *magnified*.

3. Enteromorpha intestinalis 252

    3 *a*, small portion of frond, *magnified*.

4. Calothrix confervicola (*on Ceramium rubrum*) . 284

    4 *a*, proliferous thread, *magnified*.

    4 *b*, spores, *magnified*.

    4 *c*, portion of thread, *magnified*.

XIV

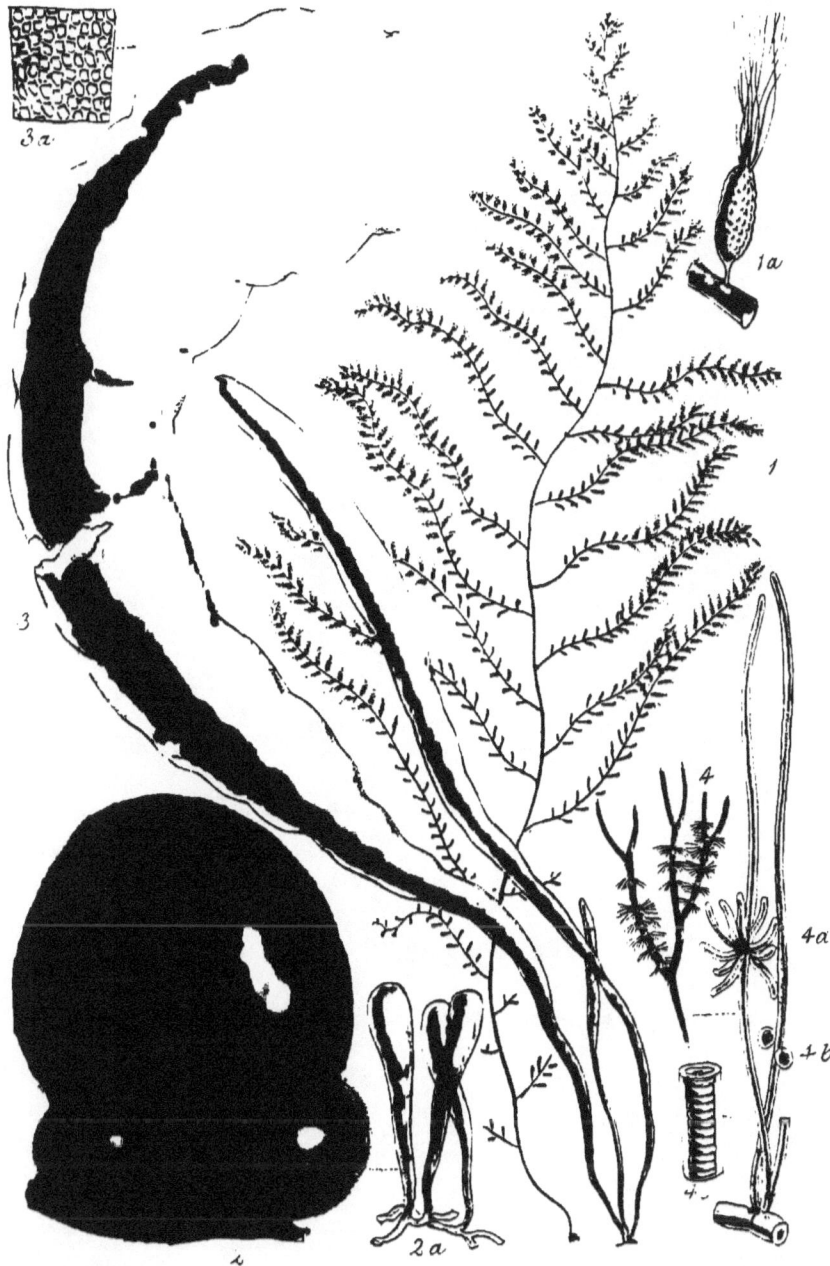

W Fitch lith. Vincent Brooks, Imp.

PLATE XV.

FIG.   PAGE

1. Striaria attenuata 62
    1 *a*, part of a branch, with spores, *magnified*.
2. Cladophora lætevirens 261
    2 *a*, part of a branch, *magnified*.
3. Cladophora lanosa 269
    3 *a*, part of a branch, *magnified*.
4. Chætomorpha Melagonium 273
    4 *a*, cells of thread, *magnified*.

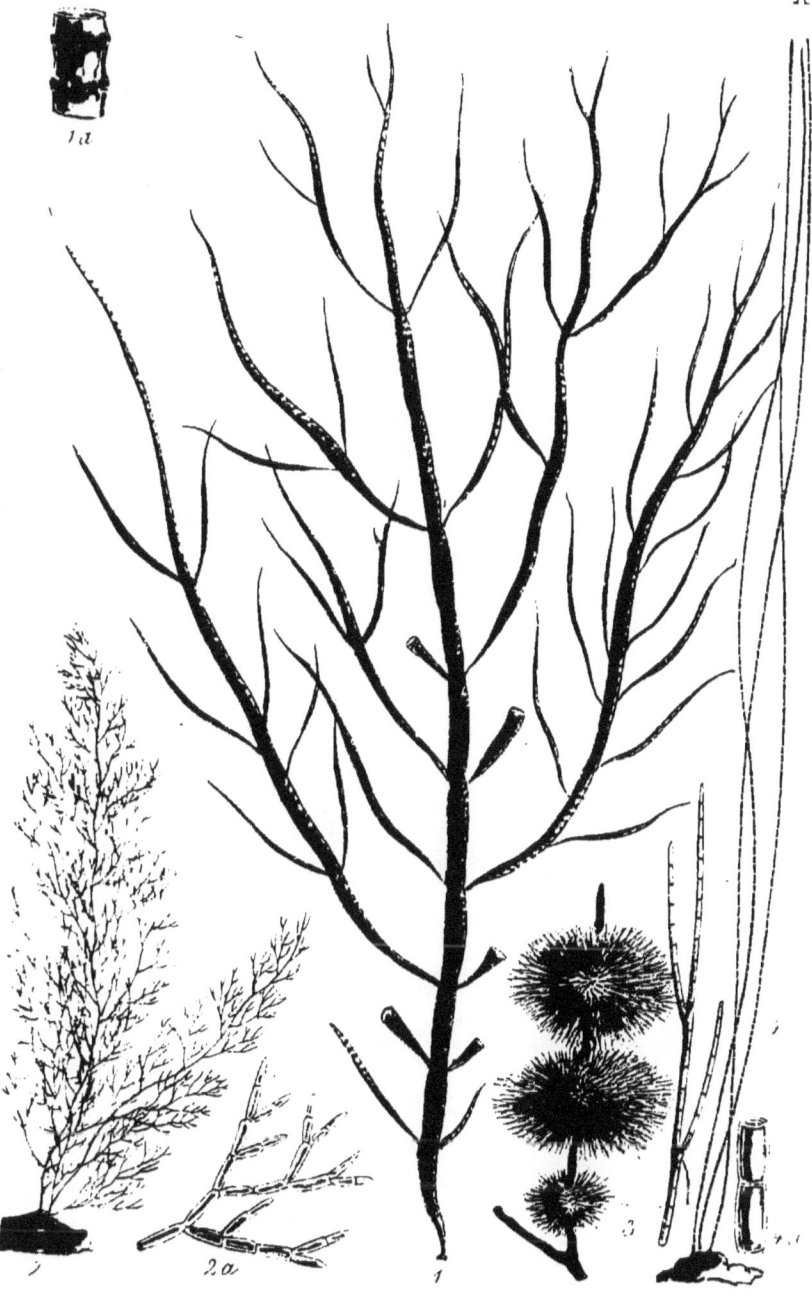

PLATE XVI.

FIG.   PAGE

1. Bryopsis plumosa 246

    1 *a*, a plumule, *magnified*.

2. Enteromorpha compressa 253

    2 *a*, part of a frond, *magnified*.

3. Cladophora falcata 263

    3 *a*, branch, *magnified*.

L. REEVE & CO.'S

PUBLICATIONS IN

Botany, Conchology, Entomology,

CHEMISTRY, TRAVELS, ANTIQUITIES,

ETC.

"None can express Thy works but he that knows them;
And none can know Thy works, which are so many
And so complete, but only he that owes them."
George Herbert.

LONDON:
L. REEVE & CO., 5, HENRIETTA STREET, COVENT GARDEN.
1867.

CONTENTS.

	PAGE
NEW SERIES OF NATURAL HISTORY	3
BOTANY	5
FERNS	11
MOSSES AND SEAWEEDS	12
FUNGI	13
SHELLS AND MOLLUSKS	14
INSECTS	16
ANTIQUARIAN	18
MISCELLANEOUS	20
RECENTLY PUBLISHED	23
FORTHCOMING WORKS	24

All Books sent post-free to any part of the United Kingdom on receipt of a remittance for the published price.
Post-Office Orders to be made payable at KING STREET, COVENT GARDEN.

LIST OF WORKS
PUBLISHED BY L. REEVE & CO.

L. REEVE AND CO.'S NEW SERIES OF NATURAL HISTORY FOR BEGINNERS.

*** A good introductory series of books on Natural History for the use of students and amateurs is still a *desideratum*. Those at present in use have been too much compiled from antiquated sources; while the figures, copied in many instances from sources equally antiquated, are far from accurate, the colouring of them having become degenerated through the adoption, for the sake of cheapness, of mechanical processes.

The present series will be entirely the result of original research carried to its most advanced point; and the figures, which will be chiefly engraved on steel, by the artist most highly renowned in each department for his technical knowledge of the subjects, will in all cases be drawn from actual specimens, and coloured separately by hand.

Each work will treat of a department of Natural History sufficiently limited in extent to admit of a satisfactory degree of completeness.

The following are now ready:—

BRITISH BEETLES; an Introduction to the Study of our Indigenous COLEOPTERA. By E. C. RYE. Crown 8vo, 16 Coloured Steel Plates, comprising Figures of nearly 100 Species, engraved from Natural Specimens, expressly for the work, by E. W. ROBINSON, and 11 Wood-Engravings of Dissections by the Author, 10s. 6d.

BRITISH BEES; an Introduction to the Study of our Natural History and Economy of the Bees indigenous to the British Isles. By W. E. SHUCKARD. Crown 8vo, 16 Coloured Steel Plates, containing nearly 100 Figures, engraved from Natural Specimens, expressly for the work, by E. W. ROBINSON, and Woodcuts of Dissections, 10s. 6d.

BRITISH SPIDERS; an Introduction to the Study of the
ARANEIDÆ found in Great Britain and Ireland. By E. F. STAVELEY.
Crown 8vo, 16 Plates, containing Coloured Figures of nearly 100 Species,
and 40 Diagrams, showing the number and position of the eyes in various
Genera, drawn expressly for the work by TUFFEN WEST, and 44 Wood-
Engravings, 10s. 6d.

BRITISH FERNS; an Introduction to the Study of the Ferns,
LYCOPODS, and EQUISETA indigenous to the British Isles. With Chapters
on the Structure, Propagation, Cultivation, Diseases, Uses, Preservation,
and Distribution of Ferns. By MARGARET PLUES. Crown 8vo, 16 Coloured
Plates, drawn expressly for the work by W. FITCH, and 55 Wood-Engravings, 10s. 6d.

BRITISH GRASSES; an Introduction to the study of the
Grasses found in the British Isles. By M. PLUES. Crown 8vo, 16
Coloured Plates, drawn expressly for the work by W. FITCH, and 100 Wood-
Engravings, 10s. 6d.

BRITISH BUTTERFLIES AND MOTHS; an Introduc-
tion to the study of our Native LEPIDOPTERA. By H. T. STAINTON.
Crown 8vo, 16 Coloured Steel Plates, containing Figures of 100 Species,
engraved from Natural Specimens expressly for the work by E. W. ROBIN-
SON, and Wood-Engravings, 10s. 6d.

BRITISH SEAWEEDS; an Introduction to the study of
the Marine ALGÆ of Great Britain and the Channel Islands. By S. O.
GRAY. Crown 8vo, 16 Coloured Plates, drawn expressly for the work by
W. FITCH, 10s. 6d.

Other Works in preparation.

BOTANY.

BRITISH WILD FLOWERS, Familiarly Described in the Four Seasons. A New Edition of 'The Field Botanist's Companion.' By THOMAS MOORE, F.L.S. One volume, Demy 8vo, 424 pp. With 24 Coloured Plates, by W. FITCH, 16s.

An elegantly-illustrated volume, intended for Beginners, describing the plants most readily gathered in our fields and hedgerows, with the progress of the seasons. Dissections of the parts of the flowers are introduced among the Figures, so that an insight may be readily obtained not only of the Species and name of each plant, but of its structure and characters of classification.

HANDBOOK OF THE BRITISH FLORA; a Description of the Flowering Plants and Ferns indigenous to, or naturalized in, the British Isles. For the Use of Beginners and Amateurs. By GEORGE BENTHAM, F.R.S., President of the Linnean Society. New Edition, Crown 8vo, 680 pp., 12s.

Distinguished for its terse and clear style of description; for the introduction of a system of Analytical Keys, which enable the student to determine the family and genus of a plant at once by the observation of its more striking characters; and for the valuable information here given for the first time of the geographical range of each species in foreign countries.

HANDBOOK OF THE BRITISH FLORA, ILLUSTRATED EDITION; a Description (with a Wood-Engraving, including dissections, of each species) of the Flowering Plants and Ferns indigenous to, or naturalized in, the British Isles. By GEORGE BENTHAM, F.R.S., President of the Linnean Society. Demy 8vo, 2 vols., 1154 pp., 1295 Wood-Engravings, from Original Drawings by W. FITCH, £3. 10s.

An illustrated edition of the foregoing Work, in which every species is accompanied by an elaborate Wood-Engraving of the Plant, with dissections of its leading structural peculiarities.

OUTLINES OF ELEMENTARY BOTANY, as Introductory to Local Floras. By GEORGE BENTHAM, F.R.S., President of the Linnean Society. Demy 8vo, pp. 45, 2s. 6d.

BRITISH GRASSES; an Introduction to the Study of the
Gramineæ of Great Britain and Ireland. By M. PLUES. Crown 8vo, 16 Coloured Plates by W. FITCH, and 100 Wood-Engravings, 10s. 6d.

One of the 'New Series of Natural History,' accurately describing all the Grasses found in the British Isles, with introductory chapters on the Structure, Cultivation, Uses, etc. A Wood-Engraving, including dissections, illustrates each Species; the Plates contain Coloured figures of 43 Species.

CURTIS'S BOTANICAL MAGAZINE, comprising the
Plants of the Royal Gardens of Kew, and of other Botanical Establishments. By Dr. J. D. HOOKER, F.R.S., Director of the Royal Gardens. Royal 8vo. Published Monthly, with 6 Plates, 3s. 6d. coloured. Vol. XXII. of the Third Series (being Vol. XCII. of the entire work) now ready, 42s. A Complete Set from the commencement may be had.

Descriptions and Drawings, beautifully coloured by hand, of newly-discovered plants suitable for cultivation in the Garden, Hothouse, or Conservatory.

THE FLORAL MAGAZINE, containing Figures and Descriptions of New Popular Garden Flowers.
By the Rev. H. HONYWOOD DOMBRAIN, A.B. Imperial 8vo. Published Monthly, with 4 Plates, 2s. 6d. coloured. Vols. I. to V., each with 64 coloured plates, £2. 2s.

Descriptions and Drawings, beautifully coloured by hand, of new varieties of Flowers raised by the nurserymen for cultivation in the Garden, Hothouse, or Conservatory.

THE TOURIST'S FLORA; a Descriptive Catalogue of the
Flowering Plants and Ferns of the British Islands, France, Germany, Switzerland, Italy, and the Italian Islands. By JOSEPH WOODS, F.L.S. Demy 8vo, 504 pp., 18s.

Designed to enable the lover of botany to determine the names of any wild plants he may meet with while journeying in our own country and the countries of the Continent most frequented by tourists. The author's aim has been to make the descriptions clear and distinct, and to comprise them within a volume of not inconvenient bulk.

A FLORA OF ULSTER, AND BOTANIST'S GUIDE
TO THE NORTH OF IRELAND. By G. DICKIE, M.D., F.L.S., Professor of Botany in the University of Aberdeen. A pocket volume, pp. 176, 3s.

A SECOND CENTURY OF ORCHIDACEOUS PLANTS,
selected from the subjects published in Curtis's 'Botanical Magazine' since the issue of the 'First Century.' Edited by JAMES BATEMAN, Esq., F.R.S. Complete in 1 Vol., royal 4to, 100 Coloured Plates, £5. 5s.

During the fifteen years that have elapsed since the publication of the 'Century of Orchidaceous Plants,' now out of print, the 'Botanical Magazine' has been the means of introducing to the public nearly two hundred of this favourite tribe of plants not hitherto described and figured, or very imperfectly so. This volume contains a selection of 100 of the most beautiful and best adapted for cultivation. The descriptions are revised and in many cases re-written, agreeably with the present more advanced state of our knowledge and experience in the cultivation of Orchidaceous plants, by Mr. Bateman, the acknowledged successor of Dr. Lindley as the leading authority in this department of botany and horticulture.

MONOGRAPH OF ODONTOGLOSSUM, a Genus of the
Vandeous Section of Orchidaceous Plants. By JAMES BATEMAN, Esq., F.R.S. Imperial folio. Parts I. to IV., each with 5 Coloured Plates, and occasional Wood Engravings, 21s.

Designed for the illustration, on an unusually magnificent scale, of the new and beautiful plants of this favoured genus of *Orchidacea*, which are being now imported from the mountain-chains of Mexico, Central America, New Granada, and Peru.

SELECT ORCHIDACEOUS PLANTS. By ROBERT
WARNER, F.R.H.S. With Notes on Culture by B. S. WILLIAMS. In Ten Parts, folio, each, with 4 Coloured Plates, 12s. 6d.; or, complete in one vol., cloth gilt, £6. 6s.

Second Series, Parts I. to III., each, with 3 Coloured Plates, 10s. 6d.

PESCATOREA. Figures of Orchidaceous Plants, chiefly
from the Collection of M. PESCATORE. Edited by M. LINDEN, with the assistance of MM. G. LUDDEMAN, J. E. PLANCHON, and M. G. REICHENBACH. Folio, 48 Coloured Plates, cloth, with morocco back, £5. 5s.

THE RHODODENDRONS OF SIKKIM-HIMALAYA;
being an Account, Botanical and Geographical, of the Rhododendrons recently discovered in the Mountains of Eastern Himalaya, from Drawings and Descriptions made on the spot, by Dr. J. D. Hooker, F.R.S. By Sir W. J. HOOKER, F.R.S. Folio, 30 Coloured Plates, £3. 16s.

Illustrations on a superb scale of the new Sikkim Rhododendrons, now being cultivated in England, accompanied by copious observations on their distribution and habits.

GENERA PLANTARUM, ad Exemplaria imprimis in Herbariis Kewensibus servata definita. By GEORGE BENTHAM, F.R.S., President of the Linnean Society, and Dr. J. D. HOOKER, F.R.S., Director of the Royal Gardens, Kew. Vol. I. Part I. pp. 454. Royal 8vo, 21s. Part II., 14s.; Part III., 15s.; or Vol. I. complete, 50s.

This important work comprehends an entire revision and reconstruction of the Genera of Plants. Unlike the famous Genera Plantarum of Eudlicher, which is now out of print, it is founded on a personal study of every genus by one or both authors. The First Vol. contains 82 Natural Orders and 2544 Genera.

FLORA OF THE ANTARCTIC ISLANDS; being Part I. of the Botany of the Antarctic Voyage of H.M. Discovery Ships 'Erebus' and 'Terror,' in the years 1839-1843. By Dr. J. D. HOOKER, F.R.S. Royal 4to. 2 vols., 574 pp., 200 Plates, £10. 15s. coloured. Published under the authority of the Lords Commissioners of the Admiralty.

The 'Flora Antarctica' illustrates the Botany of the southern districts of South America and the various Antarctic Islands, as the Falklands, Kerguelen's Land, Lord Auckland and Campbell's Island, and 1370 species are enumerated and described. The plates, which are executed by Mr. FITCH, and beautifully coloured, illustrate 870 species, including a vast number of exquisite forms of Mosses and Seaweeds.

FLORA OF NEW ZEALAND; being Part II. of the Botany of the Antarctic Voyage of H.M. Discovery Ships 'Erebus' and 'Terror,' in the years 1839-1843. By Dr. J. D. HOOKER, F.R.S. Royal 4to, 2 vols., 733 pp., 130 Plates. £16. 16s. coloured. Published under the authority of the Lords Commissioners of the Admiralty.

The 'Flora of New Zealand' contains detailed descriptions of all the plants, flowering and flowerless, of that group of Islands, collected by the Author during Sir James Ross's Antarctic Expedition; including also the collections of Cook's three voyages, Vancouver's voyages, etc., and most of them previously unpublished. The species described amount to 1767; and of the Plates, which illustrate 313 Species, many are devoted to the Mosses, Ferns, and Algæ, in which these Islands abound.

FLORA OF TASMANIA; being Part III. of the Botany of the Antarctic Voyage of H.M. Discovery Ships 'Erebus' and 'Terror,' in the years 1839-1843. By Dr. J. D. HOOKER, F.R.S. Royal 4to, 2 vols., 972 pp., 200 Plates, £17. 10s., coloured. Published under the authority of the Lords Commissioners of the Admiralty.

The 'Flora of Tasmania' describes all the Plants, flowering and flowerless, of that Island, consisting of 2203 Species, collected by the Author and others. The Plates, of which there are 200, illustrate 412 Species.

ON THE FLORA OF AUSTRALIA, its Origin, Affinities, and Distribution; being an Introductory Essay to the 'Flora of Tasmania.' By Dr. J. D. HOOKER, F.R.S. 128 pp., quarto, 10s.

HANDBOOK OF THE NEW ZEALAND FLORA; a
Systematic Description of the Native Plants of New Zealand, and the Chatham, Kermadec's, Lord Auckland's, Campbell's, and Macquarrie's Islands. By Dr. J. D. HOOKER, F.R.S. Demy 8vo. Part I., 16s.; Part II., 14s.; or complete in one vol., 30s. Published under the auspices of the Government of that colony.

A compendious account of the plants of New Zealand and outlying islands, published under the authority of the Government of that colony. The first Part contains the Flowering Plants, Ferns, and Lycopods; the Second the remaining Orders of *Cryptogamia*, or Flowerless Plants, with Index and Catalogues of Native Names and of Naturalized Plants.

FLORA AUSTRALIENSIS; a Description of the Plants of
the Australian Territory. By GEORGE BENTHAM, F.R.S., President of the Linnean Society, assisted by FERDINAND MUELLER, F.R.S., Government Botanist, Melbourne, Victoria. Demy 8vo. Vols. I., II., and III., 20s. each. Published under the auspices of the several Governments of Australia.

Of this great undertaking, the present volumes, of nearly two thousand closely-printed pages, comprise about one-half. The materials are derived not only from the vast collections of Australian plants brought to this country by various botanical travellers, and preserved in the herbaria of Kew and of the British Museum, including those hitherto unpublished of Banks and Solander, of Captain Cook's first Voyage, and of Brown in Flinders', but from the very extensive and more recently collected specimens preserved in the Government Herbarium of Melbourne, under the superintendence of Dr. Ferdinand Mueller. The descriptions are written in plain English, and are masterpieces of accuracy and clearness.

FLORA HONGKONGENSIS; a Description of the Flow-
ering Plants and Ferns of the Island of Hongkong. By GEORGE BENTHAM, P.L.S. With a Map of the Island. Demy 8vo, 550 pp., 16s. Published under the authority of Her Majesty's Secretary of State for the Colonies.

The Island of Hongkong, though occupying an area of scarcely thirty square miles, is characterized by an extraordinarily varied Flora, partaking, however, of that of South Continental China, of which comparatively little is known. The number of Species enumerated in the present volume is 1056, derived chiefly from materials collected by Mr. Hinds, Col. Champion, Dr. Hance, Dr. Harland, Mr. Wright, and Mr. Wilford.

FLORA OF THE BRITISH WEST INDIAN ISLANDS.
By Dr. GRISEBACH, F.L.S. Demy 8vo, 806 pp., 37s. 6d. Published under the auspices of the Secretary of State for the Colonies.

Containing complete systematic descriptions of the Flowering Plants and Ferns of the British West Indian Islands, accompanied by an elaborate index of reference, and a list of Colonial names.

FLORA VITIENSIS; a Description of the Plants of the Viti or Fiji Islands, with an Account of their History, Uses, and Properties. By Dr. BERTHOLD SEEMANN, F.L.S. Royal 4to, Parts I. to V. each, 10 Coloured Plates, 15s. To be completed in 10 Parts.

This work owes its origin to the Government Mission to Viti, to which the author was attached as naturalist. In addition to the specimens collected, the author has investigated all the Polynesian collections of Plants brought to this country by various botanical explorers since the voyage of Captain Cook.

CONTRIBUTIONS TO THE FLORA OF MENTONE. By J. TRAHERNE MOGGRIDGE. Royal 8vo. Parts I. and II., each, 25 Coloured Plates, 15s.

In this work a full page is devoted to the illustration of each Species, the drawings being made by the author from specimens collected by him on the spot, and they exhibit in vivid colours the beautiful aspect which many of our wild flowers assume south of the Alps.

ILLUSTRATIONS OF THE NUEVA QUINOLOGIA OF PAVON, with Observations on the Barks described. By J. E. HOWARD, F.L.S. With 27 Coloured Plates by W. FITCH. Imperial folio, half-morocco, gilt edges, £6. 6s.

A superbly-coloured volume, illustrative of the most recent researches of Pavon and his associates among the Cinchona Barks of Peru.

ILLUSTRATIONS OF SIKKIM-HIMALAYAN PLANTS, chiefly selected from Drawings made in Sikkim, under the superintendence of the late J. F. CATHCART, Esq., Bengal Civil Service. The Botanical Descriptions and Analyses by Dr. J. D. HOOKER, F.R.S. Imperial folio, 24 Coloured Plates and an Illuminated Title-page by W. FITCH, £5. 5s.

THE LONDON JOURNAL OF BOTANY. Original Papers by eminent Botanists, Letters from Botanical Travellers, etc. Vol. VII., completing the Series. Demy 8vo, 23 Plates, 30s.

JOURNAL OF BOTANY AND KEW MISCELLANY. Original Papers by eminent Botanists, Letters from Botanical Travellers, etc. .Edited by Sir W. J. HOOKER, F.R.S. Vols. IV. to IX., Demy 8vo. 12 Plates, £1. 4s. A Complete Set of 9 vols., half-calf, scarce, £10. 16s.

ICONES PLANTARUM. Figures, with brief Descriptive Characters and Remarks, of New and Rare Plants, selected from the Author's Herbarium. By Sir W. J. HOOKER, F.R.S. New Series, Vol. V. Royal 8vo, 100 plates, 31s. 6d.

FERNS.

BRITISH FERNS; an Introduction to the study of the FERNS, LYCOPODS, and EQUISETA indigenous to the British Isles. With Chapters on the Structure, Propagation, Cultivation, Diseases, Uses, Preservation, and Distribution of Ferns. By M. PLUES. Crown 8vo, 16 Coloured Plates, drawn expressly for the work by W. FITCH, and 55 Wood-Engravings, 10s. 6d.

One of the 'New Series of Natural History for Beginners,' accurately describing all the Ferns and their allies found in Britain, with a Wood-Engraving of each Species, and Coloured Figures of 32 of the most interesting, including magnified dissections showing the Venation and Fructification.

THE BRITISH FERNS; or, Coloured Figures and Descriptions, with the needful Analyses of the Fructification and Venation, of the Ferns of Great Britain and Ireland, systematically arranged. By Sir W. J. HOOKER, F.R.S. Royal 8vo, 66 Plates, £2. 2s.

The British Ferns and their allies are illustrated in this work, from the pencil of Mr. FITCH. Each Species has a Plate to itself, so that there is ample room for the details, on a magnified scale, of Fructification and Venation. The whole are delicately coloured by hand. In the letterpress an interesting account is given with each species of its geographical distribution in other countries.

GARDEN FERNS; or, Coloured Figures and Descriptions, with the needful Analyses of the Fructification and Venation, of a Selection of Exotic Ferns, adapted for Cultivation in the Garden, Hothouse, and Conservatory. By Sir W. J. HOOKER, F.R.S. Royal 8vo, 64 Plates, £2. 2s.

A companion volume to the preceding, for the use of those who take an interest in the cultivation of some of the more beautiful and remarkable varieties of Exotic Ferns. Here also each Species has a Plate to itself, and the details of Fructification and Venation are given on a magnified scale, the Drawings being from the pencil of Mr. FITCH.

FILICES EXOTICÆ; or, Coloured Figures and Description of Exotic Ferns, chiefly of such as are cultivated in the Royal Gardens of Kew. By Sir W. J. HOOKER, F.R.S. Royal 4to, 100 Plates, £6. 11s.

One of the most superbly illustrated books of Foreign Ferns that has been hitherto produced. The Species are selected both on account of their beauty of form, singular structure, and their suitableness for cultivation.

FERNY COMBES; a Ramble after Ferns in the Glens and Valleys of Devonshire. By CHARLOTTTE CHANTER. *Third Edition*. Fcp. 8vo, 8 coloured plates by FITCH, and a Map of the County, 5s.

MOSSES.

HANDBOOK OF BRITISH MOSSES, containing all that are known to be Natives of the British Isles. By the Rev. M. J. BERKELEY, M.A., F.L.S. Demy 8vo, pp. 360, 24 Coloured Plates, 21s.

A very complete Manual, comprising characters of all the species, with the circumstances of habitation of each; with special chapters on development and structure, propagation, fructification, geographical distribution, uses, and modes of collecting and preserving, followed by an extensive series of coloured illustrations, in which the essential portions of the plant are repeated, in every case on a magnified scale.

SEAWEEDS.

BRITISH SEAWEEDS; an Introduction to the Study of the Marine ALGÆ of Great Britain and the Channel Islands. By S. O. GRAY. Crown 8vo, 16 Coloured Plates, drawn expressly for the work by W. FITCH, 10s. 6d.

One of L. Reeve and Co.'s 'New Series,' briefly but accurately describing, according to the classification of the best and most recent authorities, all the Algæ found on our coasts.

PHYCOLOGIA BRITANNICA; or, History of British Seaweeds, containing Coloured Figures, Generic and Specific Characters, Synonyms and Descriptions of all the Species of Algæ inhabiting the Shores of the British Islands. By Dr. W. H. HARVEY, F.R.S. Royal 8vo, 4 vols., 765 pp., 360 Coloured Plates, £6. 6s.

This work, originally published in 1851, is still the standard work on the subject of which it treats. Each Species, excepting the minute ones, has a Plate to itself, with magnified portions of structure and fructification, the whole being printed in their natural colours, finished by hand.

PHYCOLOGIA AUSTRALICA; a History of Australian Seaweeds, comprising Coloured Figures and Descriptions of the more characteristic Marine Algæ of New South Wales, Victoria, Tasmania, South Australia and Western Australia, and a Synopsis of all known Australian Algæ. By Dr. HARVEY, F.R.S. Royal 8vo, 5 vols., 300 Coloured Plates, £7. 13s.

This beautiful work, the result of an arduous personal exploration of the shores of the Australian continent, is got up in the style of the 'Phycologia Britannica' by the same author. Each Species has a Plate to itself, with ample magnified delineations of fructification and structure, embodying a variety of most curious and remarkable forms.

NEREIS AUSTRALIS; or, Algæ of the Southern Ocean, being Figures and Descriptions of Marine Plants collected on the Shores of the Cape of Good Hope, the extratropical Australian Colonies, Tasmania, New Zealand, and the Antarctic Regions. By Dr. HARVEY, F.R.S. Imperial 8vo, 50 Coloured Plates, £2. 2s.

A selection of Fifty Species of remarkable forms of Seaweed, not included in the 'Phycologia Australica,' collected over a wider area.

FUNGI.

OUTLINES OF BRITISH FUNGOLOGY, containing Characters of above a Thousand Species of Fungi, and a Complete List of all that have been described as Natives of the British Isles. By the Rev. M. J. BERKELEY, M.A., F.L.S. Demy 8vo, 484 pp., 24 Coloured Plates, 30s.

Although entitled simply 'Outlines,' this is a good-sized volume, of nearly 500 pages, illustrated with more than 200 Figures of British Fungi, all carefully coloured by hand. Of above a thousand Species the characters are given, and a complete list of the names of all the rest.

THE ESCULENT FUNGUSES OF ENGLAND. Containing an Account of their Classical History, Uses, Characters, Development, Structure, Nutritions Properties, Modes of Cooking and Preserving, etc. By C. D. BADHAM, M.D. Second Edition. Edited by F. CURREY, F.R.S. Demy 8vo, 152 pp., 12 Coloured Plates, 12s.

A lively classical treatise, written with considerable epigrammatic humour, with the view of showing that we have upwards of 30 Species of Fungi abounding in our woods capable of affording nutritious and savoury food, but which, from ignorance or prejudice, are left to perish ungathered. "I have indeed grieved," says the Author, "when reflecting on the straitened condition of the lower orders, to see pounds of extempore beefsteaks growing on our oaks, in the shape of *Fistulina hepatica;* Puff-balls, which some have not inaptly compared to sweetbread; *Hydna,* as good as oysters; and *Agaricus deliciosus,* reminding us of tender lamb-kidney." Superior coloured Figures of the Species are given from the pencil of Mr. FITCH.

ILLUSTRATIONS OF BRITISH MYCOLOGY, comprising Figures and Descriptions of the Funguses of interest and novelty indigenous to Britain. By Mrs. T. J. HUSSEY. Royal 4to; First Series, 90 Coloured Plates, £7. 12s. 6d.; Second Series, 50 Coloured Plates, £4. 10s.

This beautifully-illustrated work is the production of a lady who, being an accomplished artist, occupied the leisure of many years in accumulating a portfolio of exquisite drawings of the more attractive forms and varieties of British Fungi. The publication was brought to an end with the 140th Plate by her sudden decease. The Figures are mostly of the natural size, carefully coloured by hand.

SHELLS AND MOLLUSKS.

ELEMENTS OF CONCHOLOGY; an Introduction to the Natural History of Shells, and of the Animals which form them. By LOVELL REEVE, F.L.S. Royal 8vo, 2 vols., 478 pp., 62 Coloured Plates, £2. 16s.

Intended as a guide to the collector of shells in arranging and naming his specimens, while at the same time inducing him to study them with reference to their once living existence, geographical distribution, and habits. Forty-six of the plates are devoted to the illustration of the genera of shells, and sixteen to shells with the living animal, all beautifully coloured by hand.

THE LAND AND FRESHWATER MOLLUSKS indigenous to, or naturalized in, the British Isles. By LOVELL REEVE, F.L.S. Crown 8vo, 295 pp., Map, and 160 Wood-Engravings, 10s. 6d.

A complete history of the British Land and Freshwater Shells, and of the Animals which form them, illustrated by Wood-Engravings of all the Species. Other features of the work are an Analytical Key, showing at a glance the natural groups of families and genera, copious Tables and a Map illustrative of geographical distribution and habits, and a chapter on the Distribution and Origin of Species.

CONCHOLOGIA ICONICA; or, Figures and Descriptions of the Shells of Mollusks, with remarks on their Affinities, Synonymy, and Geographical Distribution. By LOVELL REEVE, F.L.S. Demy 4to, published monthly in Parts, 8 Plates, carefully coloured by hand, 10s.

Of this work, comprising illustrations of Shells of the natural size, nearly 2000 Plates are published, but the plan of publication admits of the collector purchasing it at his option in portions, each of which is complete in itself. Each genus, as the work progresses, is issued separately, with Title and Index; and an Alphabetical List of the published genera, with the prices annexed, may be procured of the publishers on application. The system of nomenclature adopted is that of Lamarck, modified to meet the exigencies of later discoveries. With the name of each species is given a summary of its leading specific characters in Latin and English; then the authority for the name is quoted, accompanied by a reference to its original description; and next in order are its Synonyms. The habitat of the species is next given, accompanied, where possible, by particulars of soil, depth, or vegetation. Finally, a few general remarks are offered, calling attention to the most obvious distinguishing peculiarities of the species, with criticisms, where necessary, on the views of other writers. At the commencement of the genus some notice is taken of the animal, and the habitats of the species are worked up into a general summary of the geographical distribution of the genus.

CONCHOLOGIA ICONICA IN MONOGRAPHS.

Genera	Plates	£. s. d.	Genera	Plates	£. s. d.
Achatina	23	1 9 0	Ianthina	5	0 6 6
Achatinella	6	0 8 0	Io	3	0 4 0
Adamsiella	2	0 3 0	Isocardia	1	0 1 6
Amphidesma	7	0 9 0	Lampania	2	0 3 0
Ampullaria	28	1 15 6	Leiostraca	3	0 4 0
Anastoma	1	0 1 6	Leptopoma	8	0 10 6
Anatina	4	0 5 6	Lingula	2	0 3 0
Ancillaria	12	0 15 6	Lithodomus	5	0 6 6
Anculotus	6	0 8 0	Littorina	18	1 3 0
Anomia	8	0 10 6	Lucina	11	0 14 0
Arca	17	1 1 6	Lutraria	5	0 6 6
Argonauta	4	0 5 6	Mactra	21	1 6 6
Artemis	10	0 13 0	Malleus	3	0 4 0
Aspergillum	4	0 5 6	Mangelia	8	0 10 6
Avicula	18	1 3 0	Marginella	27	1 14 6
Buccinum	14	0 18 0	Melania	59	3 14 6
Bulimus	89	5 12 0	Melanopsis	3	0 4 0
Bullia	4	0 5 6	Melatoma	3	0 4 0
Calyptræa	8	0 10 6	Meroe	3	0 4 0
Cancellaria	18	1 3 0	Mesalia & Eglisia	1	0 1 6
Capsa	1	0 1 6	Mesodesma	4	0 5 6
Capsella	2	0 3 0	Meta	1	0 1 6
Cardita	9	0 11 6	Mitra	39	2 9 6
Cardium	22	1 8 0	Modiola	11	0 14 0
Carinaria	1	0 1 6	Monoceros	4	0 5 6
Cassidaria	1	0 1 6	Murex	37	2 7 0
Cassis	12	0 15 6	Myadora	1	0 1 6
Cerithidea	4	0 5 6	Myochama	1	0 1 6
Cerithium	20	1 5 6	Mytilus	11	0 14 0
Chama	9	0 11 6	Nassa	29	1 17 0
Chamostrea	1	0 1 6	Natica	30	1 18 0
Chiton	33	2 2 0	Nautilus	6	0 8 0
Chitonellus	1	0 1 6	Navicella & Latia	8	0 10 6
Chondropoma	11	0 14 0	Nerita	19	1 4 0
Circe	10	0 13 0	Neritina	37	2 7 0
Columbella	37	2 7 0	Niso	1	0 1 6
Concholepas	2	0 3 0	Oliva	30	1 18 0
Conus	56	3 11 0	Oniscia	1	0 1 6
Corbula	5	0 6 6	Orbicula	1	0 1 6
Crania	1	0 1 6	Ovulum	14	0 18 0
Crassatella	3	0 4 0	Paludina	11	0 14 0
Crenatula	2	0 3 0	Paludomus	3	0 4 0
Crepidula	5	0 6 6	Partula	4	0 5 6
Crucibulum	7	0 9 0	Patella	42	2 13 0
Cyclophorus	20	1 5 6	Pecten	35	2 4 6
Cyclostoma	23	1 9 0	Pectunculus	9	0 11 6
Cyclotus	9	0 11 6	Pedum	1	0 1 6
Cymbium	26	1 13 0	Perna	6	0 8 0
Cypræa	27	1 14 6	Phasianella	6	0 8 0
Cypricardia	2	0 3 0	Phorus	3	0 4 0
Cytherea	10	0 13 0	Pinna	34	2 3 0
Delphinula	5	0 6 6	Pirena	2	0 3 0
Dione	12	0 15 6	Placunanomia	3	0 4 0
Dolium	8	0 10 6	Pleurotoma	40	2 10 6
Donax	9	0 11 6	Potamides	1	0 1 6
Eburna	1	0 1 6	Psammobia	8	0 10 6
Erato	3	0 4 0	Psammotella	1	0 1 6
Eulima	6	0 8 0	Pterocera	6	0 8 0
Fasciolaria	7	0 9 0	Pterocyclos	5	0 6 6
Ficula	1	0 1 6	Purpura	13	0 16 6
Fissurella	16	1 0 6	Pyramidella	6	0 8 0
Fusus	21	1 6 6	Pyrazus	1	0 1 6
Glauconome	1	0 1 6	Pyrula	9	0 11 6
Halia	1	0 1 6	Ranella	8	0 10 6
Haliotis	17	1 1 6	Ricinula	6	0 8 0
Harpa	4	0 5 6	Rostellaria	3	0 4 6
Helix	210	13 5 0	Sanguinolaria	1	0 1 6
Hemipecten	1	0 1 6	Scarabus	3	0 4 0
Hemisinus	6	0 8 0	Sigaretus	5	0 6 6
Hinnites	1	0 1 6	Simpulopsis	2	0 3 0
Hippopus	1	0 1 6	Siphonaria	7	0 9 0

Genera	Plates	£	s.	d.	Genera	Plates	£	s.	d.
Solarium	3	0	4	0	Triton	20	1	5	6
Soletellina	4	0	5	6	Trochita	3	0	4	0
Spondylus	18	1	3	0	Trochus	16	1	0	6
Strombus	19	1	4	0	Tugonia	1	0	1	6
Struthiolaria	1	0	1	6	Turbinella	13	0	16	6
Tapes	13	0	16	6	Turbo	13	0	16	6
Telescopium	1	0	1	6	Turritella	11	0	14	0
Terebra	27	1	14	6	Tympanotonos	2	0	3	0
Terebellum	1	0	1	6	Umbrella	1	0	1	6
Terebratula & Rynchonella	11	0	14	0	Venus	26	1	13	0
Thracia	3	0	4	0	Vertagus	5	0	6	6
Tornatella	4	0	5	6	Vitrina	10	0	13	0
Tridacna	8	0	10	6	Voluta	22	1	8	0
Trigonia	1	0	1	6	Vulsella	2	0	3	0
					Zizyphinus	8	0	10	6

CONCHOLOGIA SYSTEMATICA; or, Complete System of Conchology. By LOVELL REEVE, F.L.S. Demy 4to, 2 vols. pp. 537, 300 Plates, £10. 10s. coloured.

Of this work only a few copies remain. It is a useful companion to the collector of shells, on account of the very large number of specimens figured, as many as six plates being devoted in some instances to the illustration of a single genus.

THE EDIBLE MOLLUSKS OF GREAT BRITAIN AND IRELAND, with the modes of cooking them. By M. S. LOVELL. Crown 8vo, 12 Coloured Plates, 8s. 6d.

INSECTS.

CURTIS'S BRITISH ENTOMOLOGY. Illustrations and Descriptions of the Genera of Insects found in Great Britain and Ireland, containing Coloured Figures, from nature, of the most rare and beautiful species, and, in many instances, upon the plants on which they are found. Royal 8vo, 8 vols., 770 Plates, coloured, £21.

Or in separate Monographs.

Orders	Plates	£	s.	d.	Orders	Plates	£	s.	d
Aphaniptera	2	0	2	0	Hymenoptera	125	4	0	0
Coleoptera	256	8	0	0	Lepidoptera	193	6	0	0
Dermaptera	1	0	1	0	Neuroptera	13	0	9	0
Dictyoptera	1	0	1	0	Omaloptera	6	0	4	6
Diptera	103	3	5	0	Orthoptera	5	0	4	0
Hemiptera	32	1	1	0	Strepsiptera	3	0	2	6
Homoptera	21	0	14	0	Trichoptera	9	0	6	6

'Curtis's Entomology,' which Cuvier pronounced to have "reached the ultimatum of perfection," is still the standard work on the Genera of British Insects. The Figures executed by the author himself, with wonderful minuteness and accuracy, have never been surpassed, even if equalled. The price at which the work was originally published was £43. 16s.

BRITISH BEETLES; an Introduction to the study of our
Indigenous COLEOPTERA. By E. C. RYE. Crown 8vo, 16 Coloured
Steel Plates, comprising Figures of nearly 100 Species, engraved from
Natural Specimens, expressly for the work, by E. W. ROBINSON, and 11
Wood-Engravings of Dissections by the Author, 10s. 6d.

This little work forms the first of a New Series designed to assist young persons to a more profitable, and, consequently, more pleasurable observation of Nature, by furnishing them in a familiar manner with so much of the science as they may acquire without encumbering them with more of the technicalities, so confusing and repulsive to beginners, than are necessary for their purpose. In the words of the Preface, it is "somewhat on the scheme of a *Delectus;* combining extracts from the biographies of individual objects with principles of classification and hints for obtaining further knowledge."

BRITISH BEES; an Introduction to the study of the Natural History and Economy of the Bees indigenous to the British Isles.
By W. E. SHUCKARD. Crown 8vo, 16 Coloured Steel Plates, containing nearly 100 Figures, engraved from Natural Specimens, expressly for the work, by E. W. ROBINSON, and Woodcuts of Dissections, 10s. 6d.

A companion volume to that on British Beetles, treating of the structure, geographical distribution and classification of Bees and their parasites, with lists of the species found in Britain, and an account of their habits and economy.

BRITISH SPIDERS; an Introduction to the study of the
ARANEIDÆ found in Great Britain and Ireland. By E. F. STAVELEY.
Crown 8vo, 16 Plates, containing Coloured Figures of nearly 100 Species, and 40 Diagrams, showing the number and position of the eyes in various Genera, drawn expressly for the work by TUFFEN WEST, and 44 Wood-Engravings, 10s. 6d.

One of the 'New Series of Natural History for Beginners,' and companion volume to the 'British Beetles' and 'British Bees.' It treats of the structure and classification of Spiders, and describes those found in Britain, with notes on their habits and hints for collecting and preserving.

BRITISH BUTTERFLIES AND MOTHS; an Introduction to the study of our Native LEPIDOPTERA. By H. T. STAINTON.
Crown 8vo, 16 Coloured Steel Plates, containing Figures of 100 Species, engraved from Natural Specimens expressly for the work by E. W. ROBINSON, and Wood-Engravings, 10s. 6d.

Another of the 'New Series of Natural History for Beginners and Amateurs,' treating of the structure and classification of the Lepidoptera.

INSECTA BRITANNICA; Vols. II. and III., Diptera. By
FRANCIS WALKER, F.L.S. 8vo, each, with 10 plates, 25s.

ANTIQUARIAN.

MAN'S AGE IN THE WORLD ACCORDING TO
HOLY SCRIPTURE AND SCIENCE. By an ESSEX RECTOR. Demy
8vo, 264 pp., 8s. 6d.

The Author, recognizing the established facts and inevitable deductions of Science, and believing all attempts to reconcile them with the commonly received, but erroneous, literal interpretation of Scripture, not only futile, but detrimental to the cause of Truth, seeks an interpretation of the Sacred Writings on general principles, consistent alike with their authenticity, when rightly understood, and with the exigencies of Science. He treats in successive chapters of The Flint Weapons of the Drift,—The Creation,—The Paradisiacal State,—The Genealogies,—The Deluge,—Babel and the Dispersion; and adds an Appendix of valuable information from various sources.

THE ANTIQUITY OF MAN. An Examination of Sir
Charles Lyell's recent Work. By S. R. PATTISON, F.G.S. Second Edition. 8vo, 1s.

A MANUAL OF BRITISH ARCHÆOLOGY. By
CHARLES BOUTELL, M.A. Royal 16mo, 398 pp., 20 coloured plates,
10s. 6d.

A treatise on general subjects of antiquity, written especially for the student of archæology, as a preparation for more elaborate works. Architecture, Sepulchral Monuments, Heraldry, Seals, Coins, Illuminated Manuscripts and Inscriptions, Arms and Armour, Costume and Personal Ornaments, Pottery, Porcelain and Glass, Clocks, Locks, Carvings, Mosaics, Embroidery, etc., are treated of in succession, the whole being illustrated by 20 attractive Plates of Coloured Figures of the various objects.

THE BEWICK COLLECTOR. A Descriptive Catalogue
of the Works of THOMAS and JOHN BEWICK, including Cuts, in various
states, for Books and Pamphlets, Private Gentlemen, Public Companies,
Exhibitions, Races, Newspapers, Shop Cards, Invoice Heads, Bar Bills,
Coal Certificates, Broadsides, and other miscellaneous purposes, and Wood
Blocks. With an Appendix of Portraits, Autographs, Works of Pupils, etc.
The whole described from the Originals contained in the Largest and most
Perfect Collection ever formed, and illustrated with a Hundred and Twelve
Cuts from Bewick's own Blocks. By the Rev. THOMAS HUGO, M.A., F.S.A.,
the Possessor of the Collection. Demy 8vo, pp. 562, price 21s.; imperial 8vo (limited to 100 copies), with a fine Steel Engraving of Thomas
Bewick, £2. 2s. The Portrait may be had separately, on imperial folio,
price 7s. 6d.

WHITNEY'S "CHOICE OF EMBLEMES;" a Facsimile
Reprint by Photo-lithography. With an Introductory Dissertation, Essays
Literary and Bibliographical, and Explanatory Notes. By HENRY GREEN,
M.A. Post 4to, pp. lxxxviii., 468. 72 Facsimile Plates, 42s.

A beautiful and interesting reproduction by Photo-lithography of one of the
best specimens of this curious class of literature of the sixteenth century. An
Introductory Dissertation of eighty-eight pages traces the history of Emblematic
Literature from the earliest times, and gives an Account of the Life and Writings
of Geoffrey Whitney, followed by an Index to the Mottoes, with Translations
and some Proverbial Expressions. The facsimile reproduction of the 'Emblems,
with their quaint pictorial Illustrations, occupies 230 pages. Then follow
Essays on the Subjects and Sources of the Mottoes and Devices, on Obsolete
Words in Whitney, with parallels, chiefly from Chaucer, Spenser, and Shakespeare; Biographical Notices of some other emblem-writers to whom Whitney
was indebted; Shakespeare's references to emblem-books, and to Whitney's emblems in particular; Literary and Biographical Notes explanatory of some of
Whitney's emblems, and of the persons to whom they are dedicated. Seventy-two exceedingly curious plates, reproduced in facsimile, illustrate this portion of
the work, and a copious General Index concludes the volume.

SHAKESPEARE'S SONNETS, Facsimile, by Photo-Zincography, of the First Printed edition of 1609. From the Copy in the
Library of Bridgewater House, by permission of the Right Hon. the Earl
of Ellesmere. 10s. 6d.

MISCELLANEOUS.

LIVE COALS; or, Faces from the Fire. By MISS BUDGEN, "Acheta," Author of 'Episodes of Insect Life,' etc. Dedicated, by Special Permission, to H.R.H. Field-Marshal the Duke of Cambridge. Royal 4to, 35 Original Sketches printed in colours, 42s.

The 'Episodes of Insect Life,' published in three series some years since, and so favourably received, will pleasantly recall to the mind of every one acquainted with that Work the power of graphic description, the playful fancy, the fertile imagination, and the humorous and artistic pencil of its gifted Author. The talent and genius therein displayed won, unsolicited, from the late Prince Consort, a graceful acknowledgment in the presentation to the Author of a copy of a book, 'The Natural History of Deeside,' privately printed by command of Her Majesty the Queen. The above Work comprises a series of Thirty-five highly imaginative and humorous Sketches, suggested by burning Coals and Wood, accompanied by Essays, descriptive and discursive, on :—The Imagery of Accident —The Fire in a New Light—The Fire an Exhibitor—The Fire a Sculptor.

EVERYBODY'S WEATHER-GUIDE. The Use of Meteorological Instruments clearly Explained, with Directions for Securing at any time a probable Prognostic of the Weather. By A. STEINMETZ, Esq., of the Middle Temple, Barrister-at-Law, Author of 'Sunshine and Showers,' etc. 1s.

SUNSHINE AND SHOWERS: their Influences throughout Creation. A Compendium of Popular Meteorology. By ANDREW STEINMETZ, Esq., of the Middle Temple, Barrister-at-Law, Author of 'A Manual of Weather-Casts,' etc. etc. Crown 8vo, Wood Engravings, 7s. 6d.

This Work not only treats fully all the leading topics of Meteorology, but especially of the use of the Hygrometer, for which systematic Rules are now for the first time drawn up. Among other interesting and useful subjects, are chapters on Rainfall in England and Europe in general—Wet and Dry Years—Temperature and Moisture with respect to the health of Plants and Animals—The Wonders of Evaporation—Soil Temperature—The Influence of Trees on Climate and Water Supply—The Prognostication of the Seasons and Harvest—The Characteristics and Meteorology of the Seasons—Rules of the Barometer—Rules of the Thermometer as a Weather Glass—Popular Weather-casts—Anemometry —and finally, What becomes of the Sunshine—and what becomes of the Showers.

THE REASONING POWER IN ANIMALS. By the Rev.
J. S. WATSON, M.A. 480 pp. Crown 8vo, 9s.

The object of the above treatise is to trace the evidences of the existence in the lower animals of a portion of that reason which is possessed by man. A large number of carefully-selected and well-authenticated anecdotes are adduced of various animals having displayed a degree of intelligence distinct from instinct, and called into activity by circumstances in which the latter could have been no guide.

METEORS, AEROLITES, AND FALLING STARS. By Dr. T. L. PHIPSON, F.C.S. Crown 8vo. 25 Woodcuts and Lithographic Frontispiece, 6s.

A very complete summary of Meteoric Phenomena, from the earliest to the present time, including the shower of November, 1866, as observed by the Author.

MANUAL OF CHEMICAL ANALYSIS, Qualitative and Quantitative; for the Use of Students. By Dr. HENRY M. NOAD, F.R.S. Crown 8vo, pp. 663, 109 Wood Engravings, 16s. Or, separately, Part I., 'QUALITATIVE,' 6s.; Part II., 'QUANTITATIVE,' 10s. 6d.

A Copiously-illustrated, Useful, Practical Manual of Chemical Analysis, prepared for the Use of Students by the Lecturer on Chemistry at St. George's Hospital. The illustrations consist of a series of highly-finished Wood-Engravings, chiefly of the most approved forms and varieties of apparatus.

PHOSPHORESCENCE; or, the Emission of Light by Minerals, Plants, and Animals. By Dr. T. L. PHIPSON, F.C.S. Small 8vo, 225 pp., 30 Wood Engravings and Coloured Frontispiece, 5s.

An interesting summary of the various phosphoric phenomena that have been observed in nature,—in the mineral, in the vegetable, and in the animal world.

THE ZOOLOGY OF THE VOYAGE OF H.M.S. SA-
MARANG, under the command of Captain Sir Edward Belcher, C.B., during the Years 1843–46. By Professor OWEN, Dr. J. E. GRAY, Sir J. RICHARDSON, A. ADAMS, L. REEVE, and A. WHITE. Edited by ARTHUR ADAMS, F.L.S. Royal 4to, 257 pp., 55 Plates, mostly coloured, £3. 10s.

In this work, illustrative of the new species of animals collected during the surveying expedition of H.M.S. Samarang in the Eastern Seas in the years 1843–1846, there are 7 Plates of Quadrupeds, 1 of Reptiles, 10 of Fishes, 24 of Mollusca and Shells, and 13 of Crustacea. The Mollusca, which are particularly interesting, include the anatomy of *Spirula* by Professor Owen, and a number of beautiful Figures of the living animals by Mr. Arthur Adams.

TRAVELS ON THE AMAZON AND RIO NEGRO;
with an Account of the Native Tribes, and Observations on the Climate, Geology, and Natural History of the Amazon Valley. By ALFRED R. WALLACE. Demy 8vo, 541 pp., with Map and Tinted Frontispiece, 18s.

A lively narrative of travels in one of the most interesting districts of the Southern Hemisphere, accompanied by Remarks on the Vocabularies of the Languages, by Dr. R. G. LATHAM.

A SURVEY OF THE EARLY GEOGRAPHY OF WESTERN EUROPE, as connected with the First Inhabitants of Britain, their Origin, Language, Religious Rites, and Edifices. By HENRY LAWES LONG, Esq. 8vo, 6s.

LITERARY PAPERS ON SCIENTIFIC SUBJECTS.
By the late Professor EDWARD FORBES, F.R.S., selected from his Writings in the 'Literary Gazette.' With a Portrait and Memoir. Small 8vo, 6s.

THE GEOLOGIST. A Magazine of Geology, Palæontology, and Mineralogy. Illustrated with highly finished Wood-Engravings. Edited by S. J. MACKIE, F.G.S., F.S.A. Vols. V. and VI., each, with numerous Wood-Engravings, 18s. Vol. VII., 9s.

GUIDE TO COOL-ORCHID GROWING. By JAMES BATEMAN, Esq., F.R.S., Author of 'The Orchidaceæ of Mexico and Guatemala.' Woodcuts, 1s.

THE STEREOSCOPIC MAGAZINE. A Gallery for the Stereoscope of Landscape Scenery, Architecture, Antiquities, Natural History, Rustic Character, etc. With Descriptions. 5 vols., each complete in itself and containing 50 Stereographs, £2. 2s.

THE ARTIFICIAL PRODUCTION OF FISH. By PISCARIUS. Third Edition. 1s.

RECENTLY PUBLISHED.

MOORE'S BRITISH WILD-FLOWERS. 24 Coloured Plates, 16.

GRAY'S BRITISH SEAWEEDS. 16 Coloured Plates, 10s. 6d.

STAINTON'S BRITISH BUTTERFLIES AND MOTHS. 16 Coloured Plates, 10s. 6d.

STEINMETZ'S EVERYBODY'S WEATHER-GUIDE. 1s.

PLUES'S BRITISH GRASSES. 16 Coloured Plates, 10s. 6d.

LOVELL'S EDIBLE MOLLUSKS OF BRITAIN. 8s. 6d.

STEINMETZ'S SUNSHINE AND SHOWERS. 7s. 6d.

WATSON'S REASONING POWER IN ANIMALS. 9s.

LIVE COALS; or, Faces from the Fire. By the Author of 'Episodes of Insect Life.' 42s.

PHIPSON'S METEORS, AEROLITES, AND FALLING STARS. 6s.

BENTHAM'S HANDBOOK OF THE BRITISH FLORA. New Edition, 12s.

SECOND CENTURY OF ORCHIDACEOUS PLANTS. Edited by JAMES BATEMAN, Esq. Royal 4to, Part X., with 10 Coloured Plates, 10s. 6d.; and the Work complete in 1 Vol., £5. 5s.

BATEMAN'S ODONTOGLOSSUM. Imperial Folio. Part IV., with 5 Coloured Plates, 21s.

PLUES'S BRITISH FERNS. 16 Coloured Plates, 10s. 6d.

RYE'S BRITISH BEETLES. 16 Coloured Plates, 10s. 6d.

SHUCKARD'S BRITISH BEES. 16 Coloured Plates, 10s. 6d.

STAVELEY'S BRITISH SPIDERS. 16 Coloured Plates, 10s. 6d.

BENTHAM'S FLORA AUSTRALIENSIS. Vol. III. 20s.

HOOKER'S FLORA OF NEW ZEALAND. Part II. 14s.

HUGO'S "BEWICK COLLECTOR." Demy 8vo, 21s.; Imperial 8vo, with Portrait of T. Bewick, 42s.

www.ingramcontent.com/pod-product-compliance
Lightning Source LLC
Chambersburg PA
CBHW030547300426
44111CB00009B/884